AF725965

Fireworks Displays: Explosive Entertainment
A guide to getting the most from your firework display for designers, firers and event organisers

ISBN: 978-0-8206-0064-2 (B&W Hardcover)
ISBN: 978-0-8206-0090-1 (Color Hardcover)
ISBN: 978-0-8206-0091-8 (eBook)

Chemical Publishing Company:

www.chemical-publishing.com

Printed in the United States of America

Firework Displays: Explosive Entertainment

A guide to getting the most from your firewor display for designers, firers and event organise

Dr Tom Smith

Davas Ltd

Preface

This book is designed for amateur and professional firers, designers and event organisers of display fireworks to enable them to get the most from the fireworks they use and to learn about planning events for the future. It will also be useful, we hope, to designers and event producers — if only to assist them in taking a common language to the firers!

The text has been prepared by a number of authors and the inclusion of a particular approach, technique or method by one should not be taken as endorsement by all! If you put 20 fireworkers in a room and ask them how to rig a particular item, you are apt to get 20 different, but valid, solutions!

Dedication

This book is dedicated to the memory of Mark Blanch, my cousin, who was the last of the Blanches directly involved in Brockham Bonfire, and who sadly succumbed to pancreatic cancer during the preparation of this book; and to the memory of Dr Takeo Shimizu, who taught so many people so much about the art and science of fireworks and who died at the age of 98, just before this book was completed.

Foreword

Tom Smith is to be congratulated on putting together this delightful book on the theme of firework displays. Aimed at display firers, both amateur and professional, this book covers an impressive range of theoretical and practical topics, making it unique in the fireworks literature.

Tom brings enormous experience in the fireworks industry, including a period as display manager of Kimbolton Fireworks, during which time Kimbolton won several firework competitions. More recently he has acted as a consultant for the London Millennium celebrations, the Athens Olympic Games, and the Melbourne Commonwealth Games. Together with his co-authors, Tom has assembled a truly comprehensive treatise, covering almost every aspect of firework displays, from a taxonomy of fireworks to display design, and from firework competitions to environmental impact. The chapter on risk assessment is particularly valuable, and emphasises the crucial distinction between risk and hazard, as well as adopting a very practical approach to the overall evaluation of risk.

For amateur firers the book will prove to be especially valuable. Amateurs often lack the support network, as well as the formal and informal apprenticeship opportunities, afforded to the professional. Access to the wealth of experience and information captured in this book can only lead to improved safety levels and to enhanced artistic content in amateur displays.

Beautifully illustrated, the book is also well written and a pleasure to read. Even the topic of legal issues, a subject with plenty of soporific potential, is leavened with ample personal interpretation and perspective. As a relative newcomer to the world of firework displays, I found the book to be hugely informative. This superb and comprehensive text undoubtedly has a place on the bookshelf of anyone interested in fireworks and firework displays.

Professor Chris Bishop
Vice President, The Royal Institution
Distinguished Scientist, Microsoft Research
March 2011

Acknowledgements

Firstly I would like to thank my co-authors for all the hard work and patience they have shown. Thanks are also due to Jacob, my son, for assistance in preparing the diagrams of firework types, to Ron Lancaster — for whom I worked for some 20 years as a part-time firer and then as Display Manager of Kimbolton Fireworks and who suggested the production of this book. Also to Dr Elizabeth Gilmour for her advice on medical matters and to Ben Carr and his colleagues at Chemical Publishing, my ever-patient publisher, and to David Cottingham for some inspired copyediting and corrections — any mistakes remaining are all down to me.

Finally, thanks are due to Helen my wife, Charis my daughter and Avril my secretary for their encouragement, proofreading and other helpful suggestions.

Photographs and images

Except where shown, photographs have been taken by the authors or are no longer attributable. If any reader can prove ownership of unattributed images we apologise and will be pleased to acknowledge them in subsequent editions.

Terminology

The following terms are used generally within this book without further definition.

Firing area	The area, or areas, in which the fireworks are set up
Display	The moment from the firing of the first firework to the extinguishing of the last
Fallout area	The area in which debris from the fireworks performing normally falls
"Safety" distance	The distance at which the risk to the audience or other features is reduced to an acceptable level — it is not a "safe" distance
Operator	The person or people who actually fire the firework display — whether that be in close proximity (by manual firing) or remotely (normally by electrical firing)
Audience	The people who watch a firework display — wherever they may be situated

Author notes

Tom Smith — Dr Tom Smith was introduced to fireworks by his grandfather, Jack Blanch, who was at school with and subsequently worked with members of the Brock family in the early 1900s.

Following his doctorate in chemistry from the University of Oxford, and a spell as research chemist and computer lecturer, Tom took up a full-time post as display manager with Kimbolton Fireworks in the UK, during which time the company won several firework competitions and staged some of the largest displays seen in the UK to date. In 1998 Tom left to establish an independent explosives consultancy, Davas Ltd, working worldwide on events such as the London Millennium celebrations, the Athens Olympic Games and the Melbourne Commonwealth Games as well as acting as advisor to major firework displays and explosives companies around the world. Recent firework display projects include acting as consultant to the London and Hong Kong New Year's Eve celebrations.

In addition, Tom represents the UK on a number of International Working Groups, including the development of European Standards for "professional" fireworks.

Davas Ltd now publishes the internationally recognised *Journal of Pyrotechnics* and delivers courses in Pyrotechnic Chemistry around the world.

Chris Pearce — Obtained his MSc at the University of Birmingham and, following an early career in teaching, entered the firework industry full-time in 1997. Chris is now Managing Director of Jubilee Fireworks Ltd — one of the UK's leading display companies. He was Chairman of the British Pyrotechnists Association for a five-year period and Deputy Chair of TESA (The Event Services Association). Chris contributes regularly to various academic and popular journals and oversees the BPA's Training and Examination scheme. He has been adjudicator and technical consultant for the British Musical Fireworks Championships since 2004.

Andrew Wiggins — Andy is Display Manager with Jubilee Fireworks Ltd and has been involved in the fireworks industry on a full-time basis for 18 years. Andrew is responsible for the design of all Jubilee's major shows and his credits include the "Champion of Champions" winning displays at both Plymouth (non-musical) and Southport (musical) competitions. International displays include prize-winning performances in Monaco and the Philippines, along with major shows in Malta, Spain (San Sebastian), Italy, Portugal, Macau and the prestigious Alton Towers displays in the UK.

Rodney Clarke — Rodney's pyrotechnic interests developed alongside a career in the IT industry. He was responsible for the development of firing-systems technology for Jubilee Fireworks Ltd and is now European representative for PMI (Pyrotechnics Management Inc) — manufacturers of the renowned FireOne digital firing system. Rodney's role with PMI and as an independent consultant have led to significant involvement in some of the largest firework display projects in the world, including the monumental Palm Jumeirah show in Dubai and numerous high-profile competitions and displays in Spain, Italy, France, Monaco, Portugal, Malta, Montreal and Germany (Berlin — Pyromusikale).

Andrew Walsh — Andrew Walsh AM is one of the most experienced and respected performing arts professionals in the world. With great theatrical sense, he has extensive experience creating, directing and managing large-scale events in Australia and throughout Asia, North America and Europe. In an extensive career that commenced in Melbourne in the late 1970s, Andrew has been continuously employed in the arts. He has worked for major performing arts companies, toured with international stars and directed hundreds of productions and events. He was theatrical director for Australia Pavilions at three World Expos and has directed Sydney's Australia Day spectacular for 14 years. He directed the London Millennium celebrations and in 2001 was the creative director for the widely acclaimed Centennial ceremony for Australia's Centenary of Federation.

In 2003 Andrew was Director of Ceremonies for Rugby World Cup and conceived and directed the Opening and Closing Ceremonies and pre-match segments. He was Executive Producer of the Opening and Closing Ceremonies of the 2004 Athens Olympic Games and most recently the Executive Producer and Creative Director of the 2006 Melbourne Commonwealth Games and the 2011 Athens Special Olympics.

Andrew is an associate of the Graduate School of Management at the University of Technology, Sydney and a fellow of the Australian Centre for Event Management. He was awarded the Centenary Medal for contribution to Australian Society and was appointed a Member of the Order of Australia in the 2006 Queen's birthday honours.

Darryl Fleming — Darryl has been associated with Kimbolton Fireworks for nearly 25 years and is currently the display director responsible for the creative design, choreography, delivery and production of all firework display for the company. Most notable achievements include Trafalgar 200, Dublin St Patrick's Day Skyfest, Legoland, Cowes Regatta,

Edinburgh's Hogmanay and most recently the Mayor of London's New Year's Eve celebrations on the Thames and focused on the London Eye.

Kimbolton Fireworks have had unrivalled success at many of the large International Fireworks Competitions, including Montreal, Cannes, Madeira, Barcelona, Bilbao, Croatia, Knokke and Hannover. Darryl Fleming now has a reputation as consultant for other International companies at Cannes, Montreal and Dubai.

James Donald — Jim joined Jack Morton Worldwide in 1995 and has worked his way through the organisation and is currently one of the company's leading Executive Producers. The range of event and experiential communication projects with which he has been involved has given him insights into the demands of events on all scales and for clients in all industries. These have included delivering the Mayor of London's New Year's Eve "Marking of Midnight", from 2005 through to 2011, an ambitious, high-profile, iconic event for 250,000 Londoners live and millions around the world on TV; the 2008 Olympic & Paralympics home-coming parade through London culminating in Trafalgar Square; as well as key roles within the production team for the Manchester 2002 Commonwealth Games Ceremonies and the Hong Kong Handover in 1997.

In addition to these public events, Jim has led many global experiential campaigns for corporate markets that have included Thomson Reuters, Toyota Europe, Samsung Mobile and many more. Jim's experience has given him the ability to lead complex multi-region and multifaceted projects.

Martin Smith — Martin is a director of Brockham Bonfire Limited and is responsible for the firework display. His grandfather helped re-start Bonfire in the 1930s and the family have been involved with the firework display ever since — Martin is the younger brother of the main author, Tom Smith. The Brockham Bonfire firework team have close links with Kimbolton Fireworks and regularly help with their displays both in the UK and overseas.

Murray Torrible — Murray has operated a specialist insurance facility for the Firework, Pyrotechnic and Special Effects industry since 1997. Following a hardening of the insurance market in 1996 Murray undertook 12 months of research into the operations and practices within the sector. Utilising this research, Murray identified an insurer that was willing to underwrite the risk at competitive premiums. Within two years of introducing the facility Murray was arranging the insurances for a majority of the industry sector. Murray continued to innovate and react to developments within the insurance market, developing a Captive Insurance Vehicle in 2004 to stabilise premiums for the firework sector following the increases in premiums experienced in 2002 and 2003.

Murray set up Precision Broking Ltd in 2010, a specialist industry brokerage for the Firework Pyrotechnic and Production Services Sector. The new insurance facility offers liability and property with the same insurer and a broader range of covers and limits than were previously available. Since its introduction in 2010 Precision Broking has been appointed by most of the major display operators, importers and manufacturers in the UK and is now extending its operations into Europe and the Middle East.

List of Tables

List of Figures

Table of Contents

Chapter 1 – Introduction

Tom Smith – Davas Ltd

The use of outdoor fireworks for celebration has a long-established history, although custom and practice throughout the world vary widely.

This book is intended for that wide variety of people who use fireworks in more than just a domestic way. This covers the enthusiastic amateur firer who wants to get the most from the fireworks they are legally able to purchase all the way up to the seasoned professional, although they have established techniques, favourite fireworks and favourite ways of firing, who nonetheless would like to have more information about the ways other people use fireworks to the greatest effect. This book is not intended, however, for users of pyrotechnic articles indoors, although many of the same principles apply.

Figure 1.1 – Comets fired from the London Eye and barges. Photo: Jack Morton Worldwide

As far as possible we have tried to be country-neutral in our approach, although necessarily there will be aspects that reflect the authors' experiences. The range of fireworks available to different categories of people and the specific laws applicable in each continent and country can vary widely. Furthermore, these restrictions and practices are not time-independent and we foresee great changes in the types of fireworks available in the next years for a variety of reasons, including attempts to produce worldwide harmonised standards, the possibility of existing supply routes being restricted (as well as new ones opening up) and general safety and environmental concerns over the use of fireworks at all.

This is not a reason to be unduly pessimistic, however: at every stage of the author's career there have been several instances of "the worst threat ever" to the industry and its practitioners, and we have managed to adapt and survive to date! Nevertheless, the anti-firework lobby remains strong, and their novel but often misguided attempts to ban ALL use of fireworks continues.

Hopefully this book will, in some small way, help address the concerns of this lobby, and encourage users to plan and execute displays that produce the greatest possible spectacle within budget in a way that presents the lowest risk (note, please, not "safe" — see Chapter 5) to operators, the audience, bystanders, the environment and structures.

One thing is certain — there is no absolute right way to fire a firework display. In the ten years in which the author has been the "adjudicator" of the UK's largest firework competition in Plymouth (for more information see Chapter 15) there have been some 50 companies involved and nearly 50 ways or rigging and firing the displays. Only in one case was intervention necessary to ensure the fireworks were rigged to the adjudicator's satisfaction. As each year progressed, and especially because the firing site was open to members of the industry to visit, we have noted that the quality of preparation, rigging and firing techniques and the displays themselves have improved markedly.

The common approach seems to be that people look at the display site and the display, and decide there are several things they could do better, but almost inevitably also see one or two things they recognise as being better, safer, more artistic or more novel than the way they work themselves and thus incorporate these things into their next show. In this way, over time, the whole standard is improved.

Plymouth provides this opportunity quite deliberately — it is an ideal site, beautifully positioned and secure. Sadly, because of the seasonal nature of the business (in whatever country) this sort of opportunity arises rather infrequently, mostly because of the practicalities of going to see another show when you are firing one yourself, but also because there is a degree of complacency across all sectors of the industry. "I've always done it like that" is as dangerous as it is artistically stifling — and hopefully this book will encourage users to be more creative as well as improving the safety of displays.

Chapter 2 – History of firework displays

Chris Pearce – Jubilee Fireworks

Introduction

Fireworks, in various forms, have been around for at least a thousand years. Early records indicate that fireworks and pyrotechnic devices had been developed in China by around 1000 CE. Regrettably, human nature being what it is, such devices were often used in warfare and we had to wait for another 500 years or so before firework *displays* began to appear on the scene as serious forms of public and private entertainment.

A modern firework display includes a variety of devices (shells, roman candles, rockets etc.) to produce various visual and audible effects at different heights. In a sense, it is a work of art — but the main object of any display is to *entertain* — and to that end a good display designer will make best use of the range of effects available, taking into account the natural surroundings. This has always been the case, and early displays often compensated for a limited palette of colours by using elaborate structures to enhance the firework effects.

The Italians were pre-eminent in the development of firework displays. Around 1500, fireworks were in common use for religious and cultural festivals — indeed there has always been a very strong link between religion and pyrotechnics, and this continues today (particularly in countries such as Malta). Primitive firework effects were first used in theatrical productions, essentially as scenic decorations, but gradually the fireworks themselves became a more significant feature. However, these early displays were very different from the modern aerial show, with limited effects and heavy reliance on structures to enhance the presentation.

The earliest record of a formal display in England was a substantial show staged at Warwick Castle in 1572, to celebrate the visit of Queen Elizabeth I. Her Majesty was clearly delighted by the spectacle, and other displays followed as a result — although the firework presentations would have been very different from the modern-day experience. The Warwick Castle display was perhaps the forerunner of the modern "theme show" (or the "grand theatrical displays" of the 1920s and 30s); it featured two canvas forts and two hundred performers, primitive aquatic firework effects on the river Avon, accompanied by "flaming darts", a variety of fountain effects and accompanying ordnance. The display would have been noisy and visually impressive — but with predominantly gold and silver effects; in no sense would there have been the elements of pyrotechnic design, spatial structure or colour variation that we see in contemporary fireworks shows.

Interestingly, there was a serious incident at the Warwick display of 1572, resulting in the deaths of two members of the public (Mr and Mrs Henry Cooper). A "fiery projection" landed on the unfortunate couple's house while they were asleep and they perished as a result. Of course, in those days there were no risk assessments, an almost complete disregard of safety with respect to the public and certainly no equivalent of the "Health and Safety Executive" to carry out an investigation.

Displays in the 17th and 18th centuries

Significant advances in the art and science of pyrotechnics took place in the 17th century, particularly in Europe. Italy emerged as an early fore-runner, with distinctive styles being developed in the north and south of the country. One name, that of *Ruggieri,* still survives today — now associated with the French concern *Lacroix-Ruggieri* — a company of considerable reputation. The Italian tradition was developed from religious associations, particularly the celebration of Saints' Days and adherence to the Roman Catholic faith, although the great division caused by the Reformation also manifested itself in a divergence of pyrotechnic styles! The use of manufactured structures as essential components of displays was important during this period. Known as "machines" or "temples", some of these structures were very elaborate indeed and often of great architectural merit in their own right. The "machine" would form the focal point for the display in the same sense that an elaborate vase enhances a floral presentation. Many engravings survive, illustrating in great detail how the firework effects were integrated with the machine.

In England, the leading light in pyrotechnics during the mid to latter part of the 17th century was a Swede by the name of Martin Beckman. He was responsible for the coronation displays of both James II and Charles II, several shows on the river Thames, and a number of international projects — usually associated with royal events. In common with many pyrotechnicians of his day, Beckman had a military background.

The 18th century provided the opportunity for some grand displays in Europe. Royal events in France were invariably occasions for elaborate shows, and Versailles became a prominent venue for such events. The Ruggieri family were often "imported" to design and fire these magnificent displays and, as a result, eventually took residence in France.

The Treaty of Aix-la-Chapelle, signed in 1748, resulted in widespread pyrotechnic celebrations throughout Europe. It was of great significance as it marked the end of numerous small wars and conflicts. In Paris, a planned grand display was marred as a result of quarrelling between French and Italian technicians regarding who should take precedence in firing the show. The result was a disaster, with chaotic discharging of fireworks and a "mass explosion" that led to the deaths of 40 people and over 300 injuries. Again we note the almost non-existent attention to public safety.

London had planned a show of unparalleled magnificence to celebrate the Treaty. A site was selected in Green Park and a huge "machine" was erected — 114 feet high and 410 feet long — a gigantic wooden structure, giving the appearance of a palatial façade, and constructed from timber covered with canvas. It took five months to build and was adorned with flowers, statues, classical designs and a host of elaborate features. Ruggieri, along with other Italian pyrotechnicians, was brought in to design what was envisaged to be a display of unprecedented scale. Over 10,000 individual firework effects were incorporated into the show. In addition, 100 cannon were to be fired.

The composer George Frederick Handel wrote a specially commissioned overture for the event; appropriately called *Music for the Royal Fireworks,* it was to be performed in the presence of King George II and a large royal party. In no sense was the show intended to be a "pyromusical" display in the way we understand this term today, but the event began a tradition of association between music and fireworks. The development of electrical (and particularly digital) firing in the late 20th century has bonded the two disciplines into a spectacular art-form.

Sadly, the Green Park display did not live up to expectations. A dispute arose between the English and Italian pyrotechnicians regarding the relative merits of "blackmatch", as opposed to gunpowder trains, to transfer fire between the various devices. In later chapters, we will consider the various options available to the modern display firer; in 1749 the choices were limited and an argument led to lack of attention, which in turn resulted in an explosion on-site. The north pavilion of the machine caught fire, a mishap that was eventually dealt with successfully by a makeshift fire brigade. The show itself was something of a let-down. Horace Walpole, a contemporary historian, commented that "the fireworks by no means answered the expense, the length of preparation, and the expectation that had been raised."

In particular, the wheels and major set piece devices disappointed Walpole, who observed that

> *the rockets and whatever was thrown into the air succeeded mighty well, but the wheels and all that was to compose the principal part, were pitiful and ill-conducted with no change of coloured fires and shapes.*

At the end of the performance, many fireworks were left undischarged. They were acquired by the Duke of Richmond, who gave his own display on the Thames some time later. By all accounts, this was far more successful — the firework display having a better "design" and presentation, making using of the extensive grounds of the Duke's impressive townhouse situated on the banks of the Thames at Whitehall. A musical concert was followed by discharges of rockets, water-effects from boats, a line of wheels and various gerbs (fountains) and concluding with a grand illumination of the gardens and a pavilion. This is quite the reverse of modern displays, which often commence with low-key illuminations and conclude with dramatic aerial finales.

A superb contemporary illustration (Figure 2.1) shows the whole effect of the display and also provides details of the individual fireworks. It gives

Figure 2.1 – The Duke of Richmond's fireworks

a fascinating insight into the components of a typical mid-18th-century display, with considerable emphasis on elaborate set-piece devices.

The 19th century

The latter part of the 18th century witnessed the development of "pleasure gardens" in England. These were essentially places of recreation that provided entertainment of various forms. In their early incarnations, such places offered attractions that would be considered repulsive today — bear-baiting, dog-fighting and bare-knuckle boxing were quite common — and it was not unusual to see animals such as bulls adorned with fireworks. From the 1750s onwards, many of the pleasure gardens made efforts to provide more up-market entertainment and fireworks began to play a more significant role, coupled with "illuminations" to create a pleasant ambience for a more respectable clientele. It is not surprising that many manufacturers seized the opportunities offered by these establishments to demonstrate their products and increase business.

The principal pleasure gardens in London were Ranelagh, Vauxhall, Cremorne and Bermondsey Spa. Fireworks became a regular feature at Vauxhall from 1813 and continued until the venue's closure in 1859. The closing set-piece for the final show displayed the message "Farewell for Ever". Ranelagh was the most popular and fashionable and held its first display in 1761 — a "fund-raiser" for the benefit of the Middlesex Hospital.

During this period, the name of Brock became pre-eminent in British pyrotechny. The history of Brock's Fireworks dates back to before 1720, but it was during the 19th century that the Brock family established themselves as leaders in the field. In 1865, a "Grand Competition of Pyrotechnists", the brainchild of Charles Thomas (CT) Brock, was held at the Crystal Palace, Sydenham; it was a spectacular success, with over 20,000 people attending. Thus began a series of magnificent displays at this prestigious venue, continuing until 1936. It can be argued that the first "contest" at the Crystal Palace was the forerunner of other fireworks competitions, which have, particularly in recent years, become popular once again with members of the public.

It is interesting to note the "rules" laid down for the first Crystal Palace competition. Each participant was required to display:

1. 25 coloured lights; each 2" in length and 2" diameter, 5 each of white, yellow, green, blue and red
2. 12 rockets of ½ lb calibre
3. Three tourbillions

4. 12 shells of 5" diameter
5. one set-piece
6. a finale of 200 rockets of ¼ lb calibre, 50 containing bright stars; 50 tailed stars and 100 coloured stars

Each company was allowed up to five assistants. No work could be done on-site prior to the day of the show. Use of rockets or shells of calibres greater than those specified would lead to disqualification. Even in those days there were strict "ground rules" that had to be obeyed by each competitor so that a level playing field could be established. Significant advances in chemistry during the 19th century gave rise to a greater spectrum of colours in pyrotechnics, and it is no surprise that the first fireworks competition at the Crystal Palace included a demonstration of "coloured lights" (bengal illuminations) — one of the simplest pyrotechnic devices. This would enable the judges to assess the colour purity and intensity achieved by each competitor's products.

It is interesting to note that in 1875 the Explosives Act was introduced, which laid down detailed regulations for the manufacture, sale and firing of fireworks. This was subsequently strengthened by later legislation and effectively replaced, in 2005, by the Manufacture and Storage of Explosives Regulations. Prior to 1875, regulation of fireworks-related activities had been poor although centuries earlier, in 1685, an Act of Parliament limited the manufacture, sale and display of fireworks. It was largely ignored.

The latter part of the 19th century, leading up to the First World War, was a "golden age" for firework displays. Brock's had become established as the leading lights in the UK and were contracted to provide shows throughout the world for all manner of celebrations, often associated with royal events. Other companies such as Pains and Wells also strengthened their reputations during this period, but Brock was pre-eminent. Their displays would invariably include huge and complex set-pieces; massive lancework devices depicting royal portraits, Heads of State or triumphant scenes such as battle victories were very common. This necessitated great expenditure and considerable manpower requirements, in contrast with modern displays that tend to be more economical in terms of personnel.

Even the more modest displays produced by Brock were often accompanied by a detailed descriptive programme. Many of these survive, and the following is the firing order for a display presented on August 23rd, 1894, at the Arboretum, Derby.

1. Signal maroon announcing commencement of the display
2. Illumination of the grounds by great "Crystal Palace" Lights
3. Salvo of rockets with various coloured stars, fired during the illuminations
4. Signal maroon announcing magical illumination of the scene by masses of coloured fire carefully arranged in positions chosen to produce the most beautiful prismatic effects on foliage, flowers, lawns and other salient features of the grounds
5. Device — the giant sunflower with glowing centre and a fringe of golden petals
6. Batteries of saucissons
7. Salvo of great rockets, fired in rapid succession, each with a different effect
8. Triple device — mosaic letters with intersecting lines of gold and jewels
9. Salvo of rockets with stars of every hue
10. Humorous mechanical device — the acrobat with wonderful performance on the horizontal bar
11. Flights of glowworms with jewelled heads
12. Device — the jewelled tree with brilliant foliage, multicoloured flowers and golden fruit, the apex towering aloft in a graceful spire, adorned with feathery fronds of fire, and lavishing bright jewels of every colour, flanked by revolving fountains of gold and jewels
13. Salvo of shells, 15 inches in circumference, fired in rapid succession
14. Great device — the revolving sun, 50 feet in circumference, whirling rapidly on its axis; its corona of golden fringe encircling a number of rotating wheels of prismatic light
15. Salvo of shells, 18 inches in circumference, producing posies of various flowers, fired in rapid succession
16. Great fixed device — the golden hexagon — a large piece with revolving centre of coloured fire from which radiate arms of golden light bearing wheels of coloured flame
17. Flights of whistling rockets, filling the air with wild notes of some fiery bird
18. Grand special device — the weeping willow — its foliage illustrating the changing of the seasons
19. Screen of diamond display by batteries of Brock's special Roman candles, emitting comet-like stars which leave a trail of scintillating sparks
20. Firework jugglery by batteries of Roman candles which toss into the air luminous spheres of every hue

21. The electric spreaders — Brock's greatest novelty in Roman candles, discharging stars of dazzling radiance never before attained in fireworks
22. Grand device — the Niagara of fire — falling in a golden torrent and rebounding in a bright spray of glittering sparks
23. Finale melee of shells, rockets, batteries of cobras etc.

Such programmes give a fascinating insight into the content and style of a typical public display at the turn of the century. Of particular note is the relatively small number of shells (described by circumference) and very heavy emphasis on set-piece devices — the performance of which was always described in lavish detail. The inclusion of "whistling rockets" is also a point of interest. It was not uncommon for a Brock's display to conclude with a giant waterfall effect, or similarly impressive device.

This general style of display remained common in the United Kingdom for many years and was adopted by all of the major fireworks companies until well into the 1960s. Displays were invariably hand-fired and generally more leisurely affairs; mortar tubes were usually re-loaded and substantial amounts of timber accompanied the display crew to enable erection of the set-pieces.

Some very large shows were fired in America during the late 19th century, often under the auspices of Pains or Brock's. Pains offered a range of "semi-scenic" theatrical and pyrotechnic productions, with great success — "The Fall of Babylon" being a typical example. In October 1892, Brock's fired a huge show in New York, the spectacular centre-piece of which was a grand waterfall that stretched along the entire length of the Brooklyn Bridge.

The 20th century

The early part of the 20th century witnessed more remarkable displays. Brock's further enhanced their international reputation; in 1905 the Princess of Wales visited India — accompanied by a team from Brock's who provided shows in Delhi, Indore, Mysore, and Bangalore. However, fireworks activities were naturally limited during the First World War, with manufacturers turning their attention to military pyrotechnics.

The Peace Treaty was signed in Versailles on June 29th, 1919, and on July 19th one of the UK's largest ever displays was fired in Hyde Park. The show was to include portraits of the King and Queen, images of the Great War heroes, numerous devices and set-pieces and the greatest concentration of aerial fireworks ever staged. Shells of calibre 5½ inch to 16 inch were fired in salvoes of three to fifty(!); rockets of 1 lb calibre in flights of 100; Roman candles in batteries of 200 — and a finale of 2000 rockets in a single flight. It was to be a sight to behold, but that old enemy, the weather, adversely affected the show — particularly the grand set-pieces and lancework, which did not perform flawlessly after a four-hour soaking. Despite the problems, the display was greeted with rave reviews. The report in *The Daily News* was typical:

> *The effect was a complex of sensations that it only seems possible to express by the use of too many adjectives. The show was undoubtedly vastly more marvellous than anything of the kind seen in this country before.*

This reinforces the opinion that a display can often thrill the public — even if it doesn't quite achieve what the designer set out to do.

The inter-war years were also very good for the display industry, and fireworks continued to provide high-quality entertainment for the public at seaside resorts throughout the summer. Pains shows at Cowes and other smaller regattas in the south of England proved to be very popular attractions. Brock's continued to dominate the international scene, firing shows in South Africa (eight cities as part of a royal tour), Norway and Finland — amongst others. The year 1935 was vintage, with numerous displays staged to celebrate the Silver Jubilee of King George V; over 300 displays marked the event in the UK, accompanied by 76 in various parts of Africa, eight in the West Indies, one in British Honduras, and two in British Guyana. Bombay and Karachi staged massive shows, with smaller but significant displays taking place in more exotic locations such as Aden, Mauritius and the Fiji Islands.

The second war again brought a halt to public displays for a six-year period. When hostilities ceased, Brock's was again called upon to provide a massive show to celebrate the peace. The date was set for June 8th, 1946, with the Thames being selected for the site and the location being the stretch of water between Lambeth and the Charing Cross Bridge.

The display itself was an early example of a "multimedia" show — the main aerial firework display being supported with a water-borne exhibition of illuminated fountains, rigged on 20 barges. The illumination effects were achieved by using high-intensity RAF runway projectors, fitted with coloured lenses and directed onto the water fountain jets. Twenty searchlights also contributed to the light show.

The content of the display was as follows:

- 750 aerial shells of calibres 4½", 5½", 8", 10", 12", 16" and 25" — the 4½" fired in salvoes of 50, with a single concluding 25"

- Rockets of 1 lb calibre — fired in flights of 200 (these were located on an inland area between County Hall and Charing Cross Bridge)
- Roman candles in batteries of 200
- Large mines in salvoes of 250
- 3000 aquatic devices

These effects were accompanied by two grand exhibition pieces — waterfalls spanning the entire width of the river (some 300 m), cascading from a height of 20 m, and colonnades of jewel jets displayed from two temporary bridges — one opposite the Tate Gallery and a second just below County Hall.

The Thames has provided the backdrop for many memorable displays since the Second World War. Of particular note is the 50th Anniversary of VJ Day, which was staged in 1995 and fired by *Kimbolton Fireworks.* This particular show involved five barges and some 18 tonnes Net Explosive Content (NEC) of fireworks, synchronized to music over a distance of two miles along the river. The UK's Millennium celebration display was also staged on the Thames, this time under the auspices of Australian company *Syd Howard,* working in conjunction with several British companies. More recently, the French concern *Groupe F* have presented the London New Year's Eve displays — making use of not only the Thames, but also the London Eye as a focal point for their imaginative work.

Returning to the immediate post-war period, the public's appetite for firework displays continued well into the 1950s, followed by a decline in the 1960s and 70s. Brock's, while still staging displays outside the UK (particularly in the commonwealth up to the mid-1960s), focused on the manufacture of military pyrotechnics and fireworks for the commercial market. Standard Fireworks, based in Huddersfield, had by that time become a major player — although their display work was small in comparison with their retail activities.

In the 1980s, a gradual but significant change occurred in the UK display market, largely due to relaxations in the import licensing requirements, which "opened up" the Chinese market. This gave rise to access to cheaper fireworks and led to the establishment of smaller independent outfits, whose focus was very much on displays; they were able to compete effectively with the larger more established companies. By the mid-1980s, the once mighty Brock's Fireworks was on its knees, and was purchased by Standard — but this did not give rise to a resurrection of their display division. The combined Standard-Brock operation did compete successfully in this area and began to build a reputation for itself — particularly in the provision of displays for summer proms and concerts at various stately homes, which underwent a resurgence of popularity in the 1990s. Kimbolton Fireworks, based in Cambridgeshire, and Pains Fireworks, based in Hampshire (by now under different ownership), continued to provide operator-fired displays for numerous events throughout the latter part of the 20th century. Both companies are still trading and staging major displays at home and abroad.

Fireworks competitions were re-introduced into the UK in 1997. As a result of cooperation between Plymouth City Council and the Event Services Association, the inaugural *British Fireworks Championships* took place in August of that year — the displays being fired from the Mountbatten Breakwater in Plymouth Sound (for more details see Chapter 19). The site is directly opposite Plymouth Hoe and provides an exceptional viewing aspect for members of the public. Participants in the first year were Pains Fireworks, Le Maitre Pyrotechnics, Jubilee Fireworks, the Firework Company (now *Skyburst*), Vulcan Fireworks and Fantastic Fireworks — the eventual winners. The competition is now very well established and has attracted numerous companies, from relatively small concerns to major names in the UK industry. In 2006 a special "Champion of Champions" event was held with eight previous winners competing for the coveted prize — which was won by Jubilee Fireworks with a spectacular and innovative display.

In 1999, the *British Musical Fireworks Championships* was introduced at Southport. This has proved to be very popular with the public and is staged in a more "intimate" setting than the Plymouth competition. King's Gardens, with its large marine lake, provides the backdrop for the event; the audience is relatively close and the setting is something of a natural amphitheatre. Competitors are encouraged to use the lake, and selected "aquatic" effects add an extra dimension to the shows — further enhanced by the use of music. The most successful displays at Southport are invariably those that are accurately "choreographed" to the musical sound tracks, an important aspect of modern show design that is covered in Chapter 16.

Over the past decade or so, the most significant impact on firework display design has resulted from the introduction of increasingly sophisticated digital firing systems. "Electrical firing" has been used, to a greater or lesser degree, for decades — enabling fireworks to be launched at the push of a button. In essence, early systems were all based on the principle of applying a voltage to an electrical igniter by closing a switch. "Field modules" (placed at various positions on a display site) distributed the applied voltage from a battery, via a switching panel, to numerous igniters — each of which initiated a single firework in the display. Firing wide "frontages" with multiple fireworks ignited simultaneously became a routine operation in larger displays.

System design developed rapidly post-millennium, with the incorporation of digital electronics. These systems permit individual firings to take place within a fraction of a second, at various points on a display site — previously impossible within the limitations of simple "push-button" systems (which are limited by human reaction times). Hence it is now quite straightforward to design a pyromusical display where "single-shot" Roman candles eject their stars precisely on a musical note or cue; this facilitates genuine choreography as opposed to "interpreting" the music with different types of fireworks (although this approach still has its place in display design). Recent improvements in wireless technology have been embraced by firing-system designers, now an industry in itself, and large multi-location displays can be fired with signals sent from a central console to widely separated field modules. This is developed fully in Chapter 17.

What of the future? Certainly there will be further refinements and advances in electronic firing technology. It is hard to envisage the Sydney Harbour New Year's Eve Display or major competition shows being undertaken without the assistance of a complex digital firing system. No doubt there will be further improvements in pyrochemistry and the development of "cleaner" effects with lower smoke emission. The only limitations are those of human imagination and ingenuity — but as the boundaries are pushed even farther, it is comforting to note that the basic ingredients of a good firework display remain essentially the same as they have for centuries.

Chapter 3 – Firework types and effects

Tom Smith - Davas Ltd

The variety of fireworks types available to the amateur and professional users alike is quite diverse, and trying to rationalise this diversity is a daunting task. The following list of types has been developed from the recently developed EU Standard for Fireworks[1] and reflects the relevant Working Group's desire to maintain as great a flexibility for professional users as possible while developing type-specific testing methods, labelling and performance criteria.

At the time of development of the Standards, the Working Group believed that all types of Category 4 fireworks (those for use by "persons with specialist knowledge") are covered, but recognised that other types may also be available for use but are included in other standards. For instance, sparklers and throwdowns are covered by the Category 1, 2 and 3 Standards, while items that could reasonably be considered as theatrical items (e.g., jets and line rockets) are covered by T1 and T2 Standards.

Table 3.1 – Comparison of European and British Standard BS 7114

<table>
<tr><th>Category</th><th>Effective old BS 7114 category</th><th>Comments</th></tr>
<tr><td>Category 1 firework</td><td>Category 1 firework (indoor)</td><td rowspan="4">NB: The requirements of the Standards have changed from BS 7114.

There is a period (until 4 July 2017) in which BS 7114 items may continue to be supplied within the UK.</td></tr>
<tr><td>Category 2 firework</td><td>Category 2 firework (garden)</td></tr>
<tr><td>Category 3 firework</td><td>Category 3 firework (display)</td></tr>
<tr><td>Category 4 firework</td><td>Category 4 firework (display)</td></tr>
<tr><td>T1 Pyrotechnic article</td><td>None</td><td rowspan="2">For theatrical use

T1 is subdivided into T1 — indoor and outdoor, and T1 — outdoor only</td></tr>
<tr><td>T2 Pyrotechnic article</td><td>None</td></tr>
<tr><td>P1 Pyrotechnic article</td><td>None</td><td rowspan="2">Including pyrotechnic articles for vehicles</td></tr>
<tr><td>P2 Pyrotechnic article</td><td>None</td></tr>
</table>

The following types and subtypes are defined in the Category 1, 2, 3 and 4 European Firework Standards. Where necessary, additional information is given to clarify or expand the formal definitions developed. The original definitions (slightly adapted here) are in *italics* for clarity.

Aerial wheel

An aerial wheel consists of tubes containing propellant charges and sparks-, flame- and/or noise-producing pyrotechnic compositions, the tubes being fixed to a supporting structure, designed to rotate and ascend into the air. Some of the tubes (if not all) are fixed in such a way that the device ascends, in an unsupported manner, into the air by the action of the devices. The principal effect is rotation and ascent, with emission of sparks and flames, producing a visual and/or aural effect in the air.

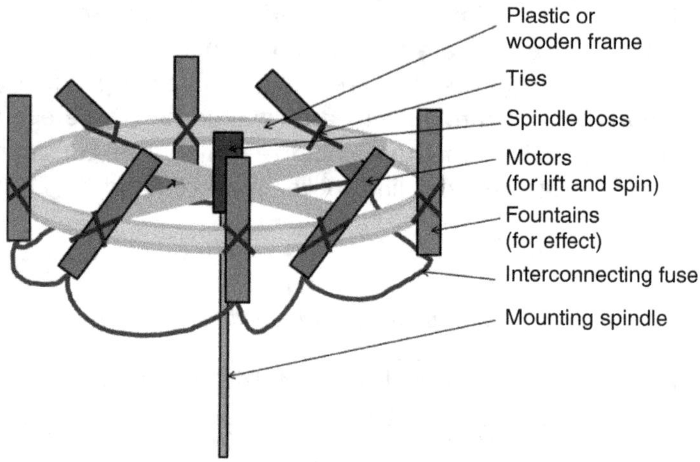

Figure 3.1 – Aerial or "crown" wheel

Such aerial wheels can be "single acting" (where the piece ascends only once) or "double acting" (where the device ascends, then drops, and then ascends again — often to the bemusement and delight of the audience). As a result there are several possible designs of this type of device, and the above illustration should only be taken as indicative.

Typically the device functions by first lighting the drivers (gerbs whose main function is thrust rather than visual effects), which rotate the wheel on its axis — to impart gyroscopic stability to the wheel, and then the rising motors function to propel the device into the air.

The major safety issue with aerial wheels is that the device is very susceptible to the effects of wind and can travel significant distances from the point of firing. Furthermore, if all the drivers do not function correctly the device can become unstable and erratic in flight.

Aqua firework (aquatic firework, nautical firework)

A firework designed to be floated on or near the surface of water by means of a buoyancy device or by itself and to function on or below water. The same effects are produced as for Bengal flames, fountains, mines, shells etc.

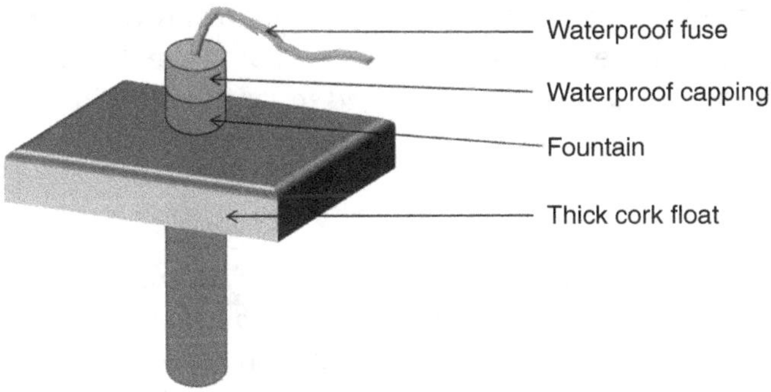

Figure 3.2 – Water fountain mounted on cork float

Subtypes include:

Underwater fireworks/sub-aquatic fireworks — *Fireworks designed to function under the water near the surface. These articles have the capacity to float on or at a few centimetres under the surface of water.*

Aquatic fireworks can be pre-rigged in their eventual firing position, or be propelled into their firing position either manually (for instance the

Figure 3.3 – Water shell bursting

hand-thrown water gerb illustrated in Figure 3.2) or from a mortar in the same way as a mine (for multiple effects) or a shell in mortar (for a single effect). In the latter cases the mortars are fired, typically at 45°; lifting charges used are usually considerably less than the equivalent charge if the device were fired vertically upwards.

The major safety concern of propelled aquatic fireworks is the range that they can achieve and the variation in that range depending on the exact conditions in which they are fired (quantity of lifting charge, the mortar used, the angle of the mortar and its elevation).

Combination

A Combination device is an assembly including several elements, of one or more types, each corresponding to one of the individual types of firework listed in this section, with one or more points of ignition.

The individual fireworks may be fused together in series or parallel, with or without delay fuses, to give their effects in a sequence or at the same time. The combination is not necessarily made at the manufacturing level and can be achieved at the display's firing place.

Compound fireworks (in which several effects are within the same single-tubed article — e.g., a Roman candle) are not considered as combinations.

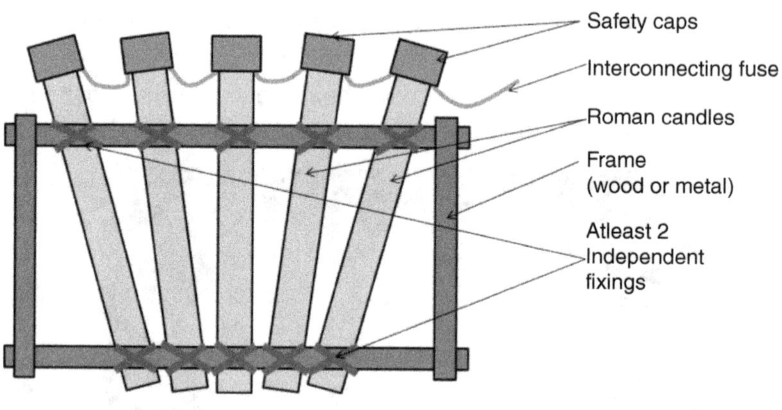

Figure 3.4 – Battery of Roman candles

Subtypes of combinations include:

Battery *— An assembly including several elements, each of the* ***same*** *type and corresponding to one of the types of firework listed, with one or more points of ignition. (This definition is included as it is used historically in category 1, 2 and 3 fireworks.)*

Figure 3.4 is typical for a Roman candle battery containing candles of the same calibre. However, more complex combinations may also be constructed containing Roman candles of various calibres and other firework types. The distinction between a "combination" and a "battery" is somewhat artificial, and serves little purpose — if there is more than one individual firework on an assembly, then the function of the whole device and the risks posed from that whole device are related to the performance and risks of the individual components.

***Cake** — An assembly including several elements either containing the same type or several types in which the initial fuse transmits fire from one tube to the next to fire the devices sequentially or in some other pattern. This article differs from a general battery or combination by the fact it is fully integrated in a unique pre-programmed product, at the manufacturing level, then cannot be dissociated.*

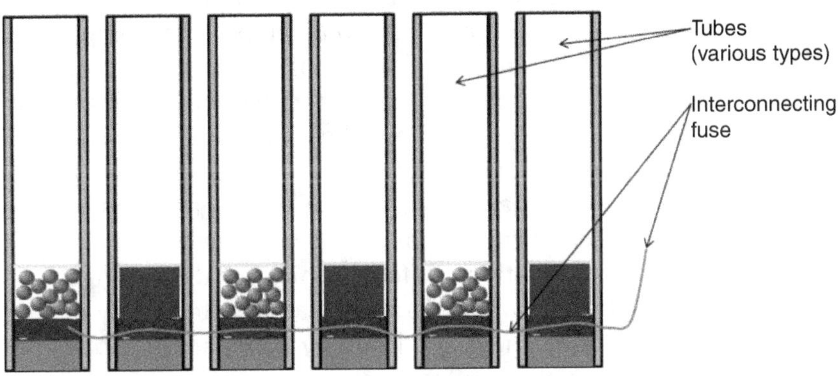

Figure 3.5 – Schematic cross-section of a typical "cake"

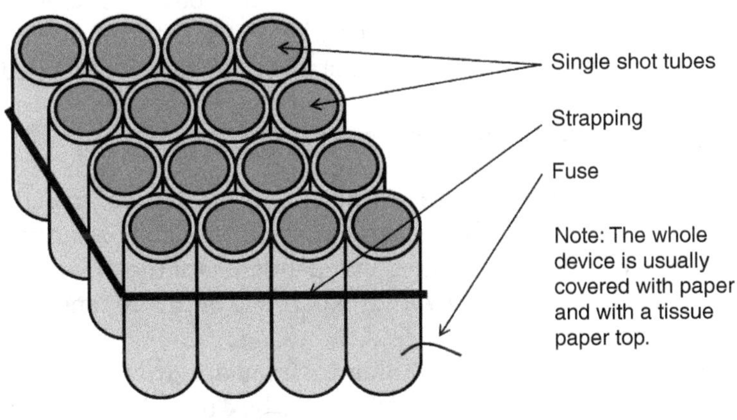

Figure 3.6 – Schematic of 4 × 4 16-shot "cake"

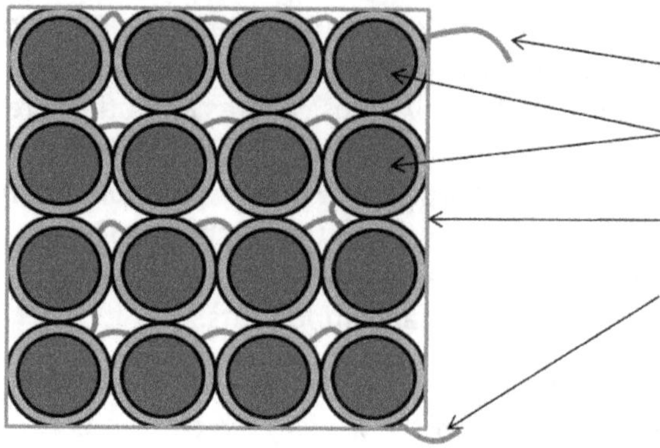

Figure 3.7 – Schematic of 4 × 4 16-shot "cake" showing internal fusing

The term "cake" originates from early examples of this type of firework that were produced in China and resemble a typical cylindrical form of the eponymous confectionary! Modern "cakes" are considerably more complex and varied in construction. In recent years a variety of complex cakes producing "chase"-type effects (see Chapter 11) have been developed and are variously described as, for instance, "Z" cakes (where the chase is from one side to the other, back again and repeated). However, this pattern is sometimes not obvious from the external appearance of the device, and care should be taken to ensure that such a device is not used inappropriately.

Given the differing orientations, different timings and different effects available within each tube, there are almost an infinite variety of cake types available, and often the name given by manufacturers does not adequately describe the actual effect observed.

It is essential that cakes are adequately supported in their firing. Cakes, even those that are heavy and with a large base and hence a low centre of gravity, rock during firing. Accidental tipping, or disruption of the cake during firing, leads to the possibility of firing in unintended and undesirable directions.

The two main safety issues with cakes are:

- That the device fails part way through due to poor construction or because of dampness. Often cakes are fitted with an auxiliary fuse to allow the finale shots to be fired independently of the main fuse. If a failure occurs, then dealing with the partially fired item and disposal of it is a serious problem.
- Sometimes the cake may continue to smoulder after firing is complete, or indeed fire may spread to the box used to transport the device (which is usually not removed completely before firing).

Lancework *– An assembly of lances on a frame, which are fused together to one or more points of ignition and are fired simultaneously or in sequence to produce bright, multi-coloured, dotted decorative patterns. Lanceworks may include fixed and mobile parts. In the latter case, tubes containing propellant charge are used to give motion to the mobile parts.*

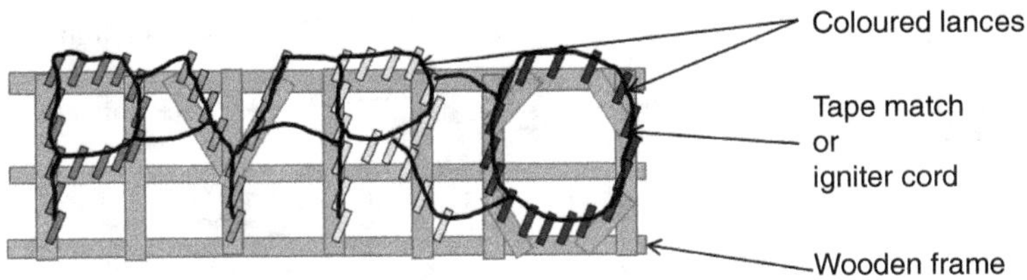

Figure 3.8 – Schematic of lancework device

Lancework can be exceedingly complex, particularly where a sponsor demands their corporate logo be faithfully reproduced (and often they cannot accept that either it should be reversed in colours — black lancework is not very impressive — or that the firework company cannot reproduce the Pantone© colour exactly!). In essence the way lancework works is a trick of the eye — the brain "joins the dots" to produce the desired pattern. It is possible to over-complicate lancework, which leads to the eye failing to make the right connections, as well as inevitably increasing the amount of smoke produced — which also leads to a diminution of the effect.

The lances are connected together by a quick-burning, preferably waterproof fuse. Increasingly this takes the form of "tapematch", a self-adhesive tape with a granulated blackpowder core which is simply folded over each lance and between each lance to form a tube, and which burns exceedingly rapidly. The disadvantage of tapematch, apart from whether it is usually classified properly, is that during the display it is extremely sensitive to accidental ignition from stray sparks.

The latticework frame construction is critical — mounting lances on a solid frame does not allow the wind (which is present at almost all displays!) to dissipate the smoke produced, and the resulting effect is ruined.

Set-piece *– An assembly including one or multiple elements which is designed not to rotate. Generally these elements belong to the family of lances, fountains and cascades, but can also include bangers, ground maroons and/or whistles.*

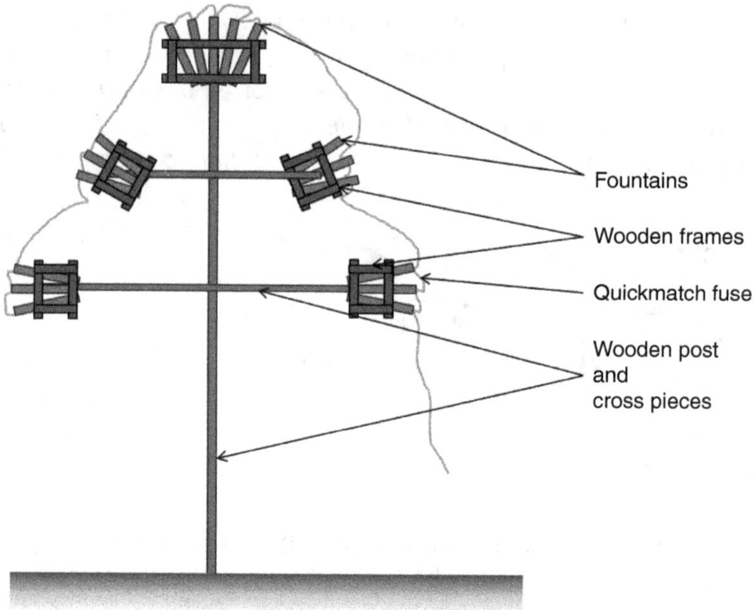

Figure 3.9 – Set-piece made from fountain bouquets

Set-pieces are usually mounted on poles.

Wheel *– An assembly including a tube or tubes containing pyrotechnic composition and provided with a means of attaching it to a support so that it can rotate.*

Set-pieces can be simple, static arrangements of gerbs (fountains), or complex moving structures producing intricate patterns and varieties of colours. Purists feel that any motive force to drive moving set-pieces should be derived from the fireworks themselves. ("Drivers" are a special type of gerb that generally produce more thrust but less visual effect than the equivalent gerb.)

Component articles

These are usually included in other fireworks and generally are without lifting charge, and thus require further preparation. See below for an indicative list of such components.

Whistle *– A tube containing pressed, whistling pyrotechnic composition, with or without sparks, with or without report-producing pyrotechnic composition.*

Figure 3.10 – Maltese complex wheel

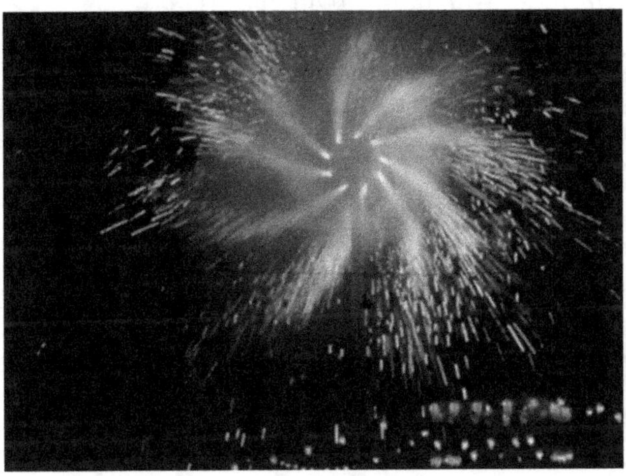

Figure 3.11 – Simple wheel in action

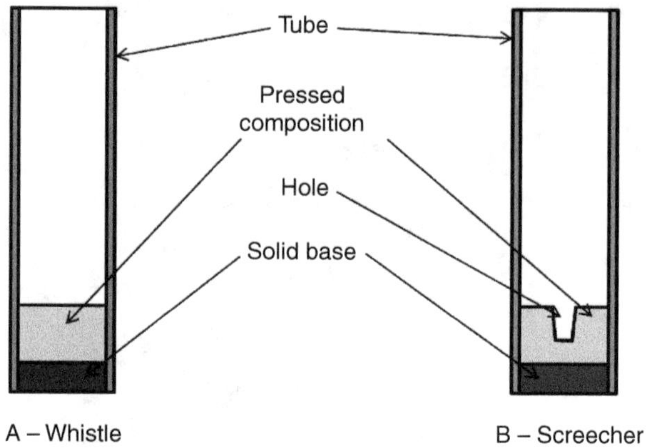

Figure 3.12 – Cross-section of A) whistle and B) screecher units

The mechanism of whistle production has been the subject of much debate,[2] but in the simplest terms it can be considered like an organ pipe, where the pressure-induced increase in burn rate, and hence gas production at the burning surface, reinforces the sound wave in the tube above the burning composition. A "screecher" is similar in construction to a simple whistle but produces a more complex rasping sound, created by interference patterns arising from sound waves produced from a central hole within the pressed composition.

Hummer *– A tube containing pyrotechnic composition designed to burn in such a way that the tube rotates in the air, producing a constant or variable pitch "note" with or without a final report.*

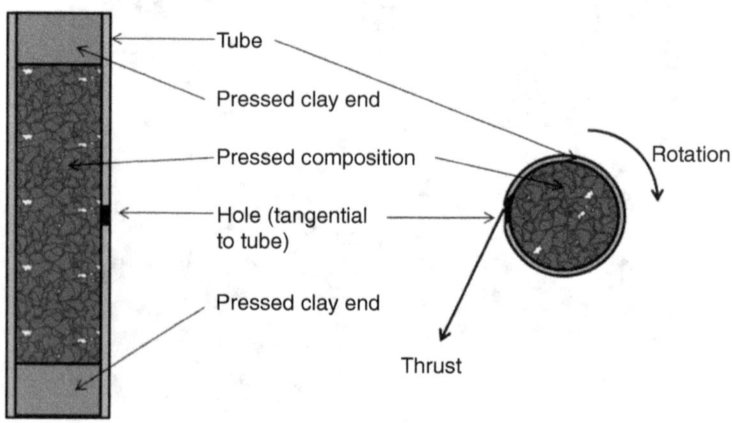

Figure 3.13 – Hummer unit

The hummer unit comprises a tube in which a hole has been formed at a tangent to the tube, causing the tube to rotate on the production of gas from the burning composition. Similar effects, whose primary effect is not noise, can be made with coloured compositions and manifest themselves as whirling, spiralling "go-getters".

Star *– A small element of compacted pyrotechnic composition, intended to burn in the air and give an individual visual effect. Stars may have various shapes: spherical, cylindrical, cubic, rectangular etc. and be obtained through various industrial processes: pressed, rolled, extruded etc. Their surface may be partially covered by combustion inhibitors or not. They can include a pyrotechnic charge to break them into fragments during their combustion to improve or modify the visual effect ("fragmentation" stars).*

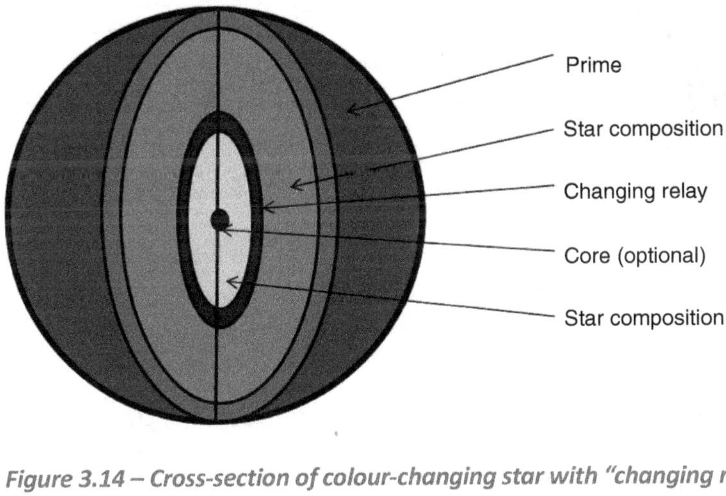

Figure 3.14 – Cross-section of colour-changing star with "changing relay"

Stars can also be pressed in moulds to form highly consistent, accurate burning cylinders of, usually, blackpowder-based compositions.

Comet *– A pyrotechnic component of a firework (e.g., Roman candle or a shot tube), containing a single solid pyrotechnic composition, which leaves a trace as it ascends up to the apex of its flight.*

Comets may form parts of shells (e.g., "spider shells") where a geometric radial burst of a few "thick" stars contrasts with the more usual spherical burst of typical colour shells, and sometimes the two effects may be combined. Comets are typically made by pressing composition into a mould rather than by building up layers onto a core as in a round star. Such pressed stars can be more consistent in performance than rolled stars as well as being easier to produce in mechanised production lines, but are generally restricted to compositions based on blackpowder. We expect the use of pressed stars to increase in the future.

Increasingly comets are used in "single-shot" Roman candles, used particularly when fired from structures — see Chapter 11.

Crossette *– A tailed star that breaks abruptly into burning fragments exhibiting a cross-shaped expansion with tail effects.*

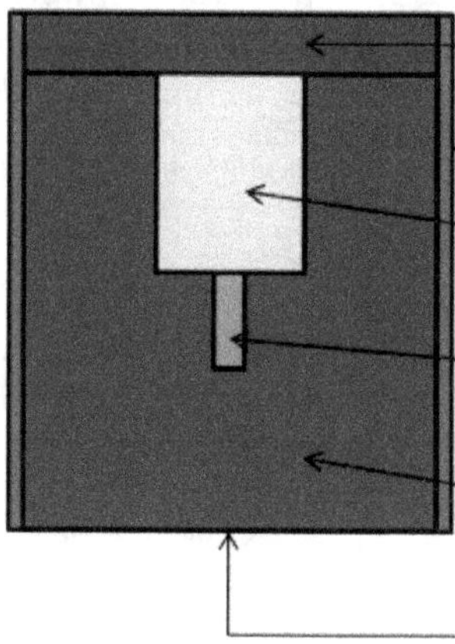

Figure 3.15 – Crossette or "Splitting Comet"

Motor *– An article consisting of a charge of explosive, generally a solid propellant either contained or not in a cylinder fitted with one or more nozzles.*

Typically motors are used to propel rockets into the air, but small rocket-like items also may be used as payloads in shells to produce an accelerating effect on burst.

Fuse *– Often a small tube or cord containing a pressed or compacted pyrotechnic composition that burns in a cigarette way and delivers a thermal effect to ignite a pyrotechnic charge located downstream. By extension, this term also applies to other types of fire transmission devices like quickmatch or blackmatch or pressed fuse.*

This component includes a **delay fuse** — a fuse incorporated into the initial fuse of a firework to introduce a delay between firing and functioning OR the internal fuse in a firework to enable sequential firing of elements of the firework (e.g., in a shell). It is unfortunate that the same term is frequently used for these two disparate components of a firework — but usually the meaning is clear in context.

One peculiar use of small pieces of slow-burning fuse is as "fish" in shell bursts. Here small lengths of fuse are packed into a weakly burst shell; as each one lights, a small amount of light (usually gold or silver) is produced with a small amount of thrust. This, combined with the low mass and density of these fuse lengths, means they fall to earth slowly and "swim" as they fall.

Butterfly *– A firework component comprising a tube that is burst, spreading stars or shining effects from both ends.*

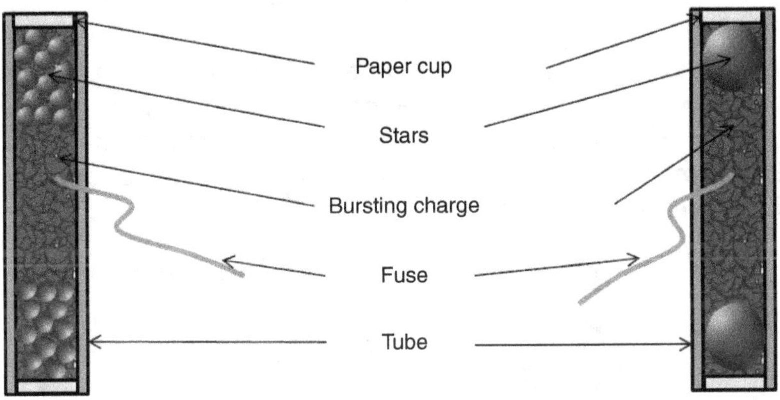

Figure 3.16 – Two types of butterfly effect

Cracker *– A firework component that burns with a crackling sound and may optionally emit a glittering or sparkling effect.*

More generally crackers produce a small "bang" and should, we feel, be distinguished from true crackling stars, which produce a column of small bangs as they burn to completion while travelling through the air.

Bombette *– A pyrotechnic component of a firework (e.g., Roman candle or a shot tube), similar to a small shell, which may optionally leave a trace as it ascends, and which bursts at or near the apex of its flight.*

Fountain

A case containing sparks and/or flame-producing pyrotechnic composition and designed to be placed on the ground, or to be fixed in the

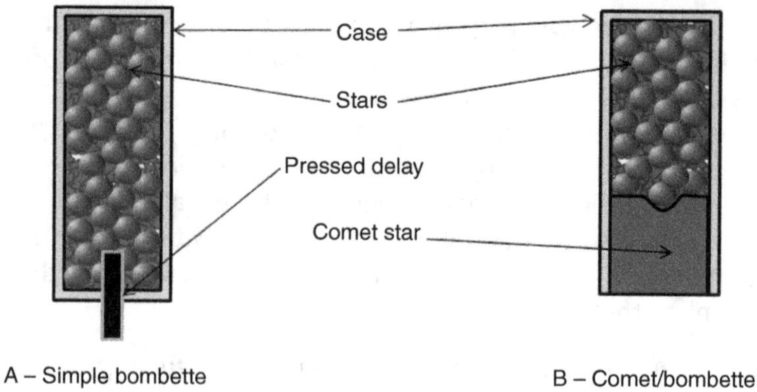

Figure 3.17 – Bombette types

ground, or to be fixed to a support, or to be held in the hand. The pyrotechnic composition may be pressed or not in the tube, with or without a choke or other constriction incorporated into the tube. The principal effect is emission of sparks and flames with aural effect other than report or without any aural effect.

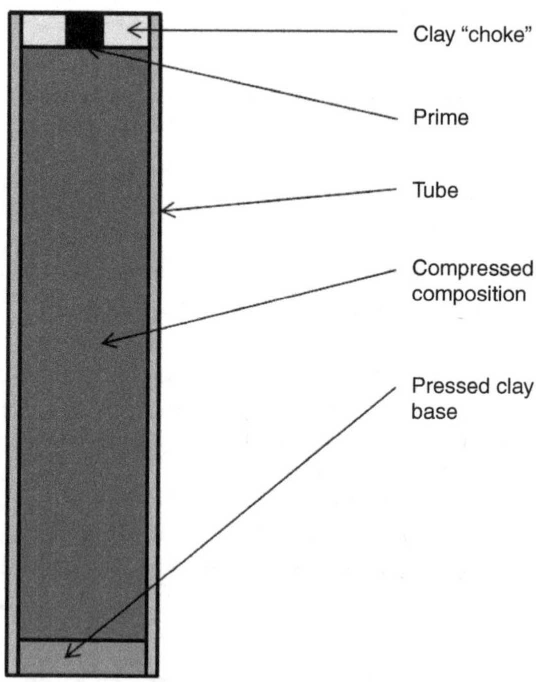

Figure 3.18 – Choked fountain

Gerb – Another name for fountain.

Subtypes of fountain include:

Bengal flame *– A tube containing a slow-burning pyrotechnic composition that may or may not be pressed. The tube has no choke and optionally burns away during functioning.*

Saxon *– A tube intended to be attached to a support in its middle so that it can rotate, and containing one or two pyrotechnic compacted charges that burn on opposite sides and eject their combustion products sideways so that rotation is obtained.*

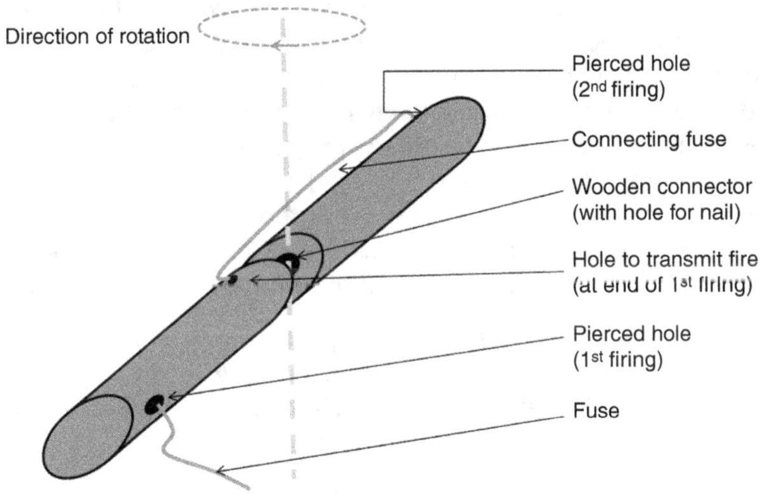

Figure 3.19 – Saxon

Lance *– A small-diameter tube containing a compacted pyrotechnic composition, burning in a cigarette way, intended to deliver a thermal output to ignite manually other fireworks or a small white or coloured flame to be used in "lancework". The pyrotechnic composition may be pressed or simply consolidated. The tube has no choke and generally burns away during functioning.*

Portfire *– A hand-held device containing a slow-burning pyrotechnic composition and emitting a small flame. A portfire can be considered a hand-held lance, as is the case when it is used to ignite manually other fireworks.*

Driver *– A device intended to produce thrust, often with limited visual effect, to cause, for instance, rotation of a wheel.*

***Strobe** – A tube containing an intermittently burning pyrotechnic composition to produce long and rapid series of flashes at a relatively constant frequency. The pyrotechnic composition may be pressed or not. The tube has no choke and optionally burns away during functioning with the emission of series of flashes.*

The burning mechanism for a strobe is not well understood, although there are several theories[3] published for this peculiar rhythmic burning of what is, essentially, a homogeneous composition.

The author believes that the glitter[4] and strobe effects are both manifestations of the same basic chemical processes, which lead to "flash" reactions occurring away from the burning surface (in the case of glitter) and at the burning surface (for strobes). In each case the proposed mechanisms for the observed effect require atmospheric oxygen to oxidise intermediate reaction by-products. Research is ongoing to explain these phenomena more fully. However, whereas glitter effects are somewhat limited in chemical composition, there are a wide variety of compositions that exhibit unstable, strobe-like burning, however counter-intuitive this may be. Pyrochemical reactions, especially those involving several steps that may or may not require the presence of atmospheric oxygen, are almost certainly not mechanisms similar to oscillating reactions in solution.[5]

***Tourbillion** – A tube or tubes containing pyrotechnic composition, which burns in a way that gives a rotary motion to the tube. This article differs from a spinner by its design — a tourbillion has no aerofoils, and functions by the lateral ejection of the combustion products.*

***Waterfall** – A case containing pressed or consolidated pyrotechnic composition that produces sparks and flame and generally consumes the tube while burning. Combustion products are ejected from the flame zone at low speed, then drop downwards like water in a waterfall.*

***Volcano** – A conical device containing consolidated or pressed composition in which the effect (height or intensity) increases as the device burns.*

As the conical fountain burns, the surface area of burning increases but the hole at the top of the cone also increases — the end result being that the height of burning remains approximately constant but the intensity of the effect increases.

Guided firework (line rocket)

An article containing pyrotechnic composition equipped to function along a rope or other guide and to produce a visual and/or aural effect.

Figure 3.20 – Waterfall effect — Japan

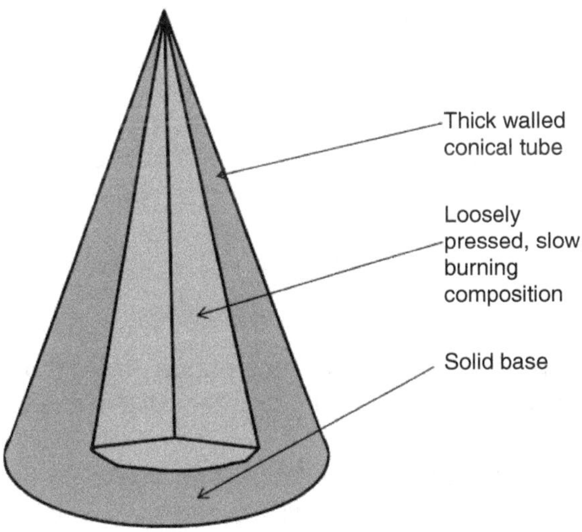

Figure 3.21 – Schematic of conical fountain (volcano)

For example, line rockets can also be used to put fire to other fireworks, generally fixed on a frame located at a distance from the firing place, e.g., at the top of a steeple or a tower. The principal effect is emission of a visual and/or aural effect.

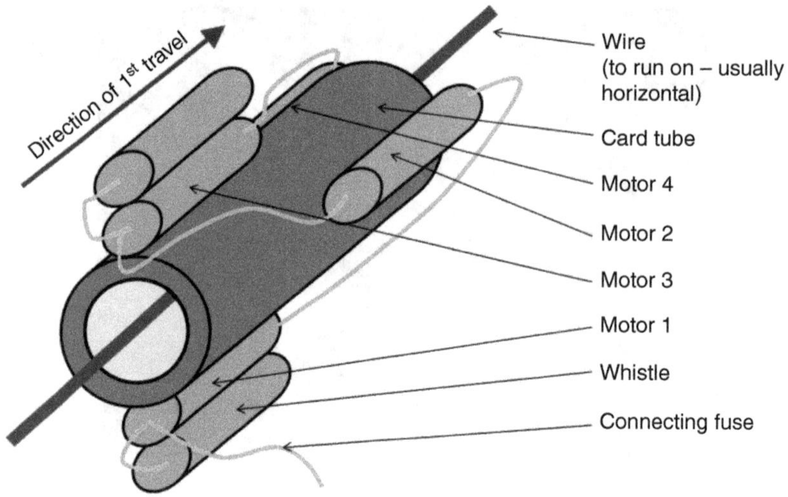

Figure 3.22 – Schematic of "Flying Pigeon". Note that the pigeon normally travels horizontally — it is shown at an angle for clarity.

Subtypes of guided fireworks include:

Pigeon *– A device that travels sequentially in both directions along the supporting guide.*

Line Rockets – more usually used indoors — these are single-acting devices.

Brock's Fireworks used to market a combination device that they termed "rats" — a series of single-acting line rockets that followed each other (scurried) along the line.

The author recalls rigging complex pigeon fireworks on wires hung above the crowd at open-air displays, initially leading to some concern, if not panic, that a firework was firing over their heads out of control. Only when the pigeon returned along the wire and then repeated its performance were the crowd reassured. Sadly, such "irresponsible behaviour" would hardly be tolerated in our risk-averse modern blame culture! Substituting wire for the normal rope supplied with a pigeon increases the potential range significantly — but care must be taken that the pigeon does not destroy itself when hitting the support structure for the wire at the end of its first flight. Spacers are often employed to prevent the pigeon hitting the support.

Mine

An article that may include integral mortar, containing propellant charge and more than one pyrotechnic unit, having as main effect the discharge of all the pyrotechnic units in a single ejection. The pyrotechnic units can be stars, bangers, butterflies, crackers, hummers, spinners/tourbillions, whistles, etc. The principal effect is ejection of all the pyrotechnic units in a single burst producing a widely dispersed visual and/or aural effect in the air.

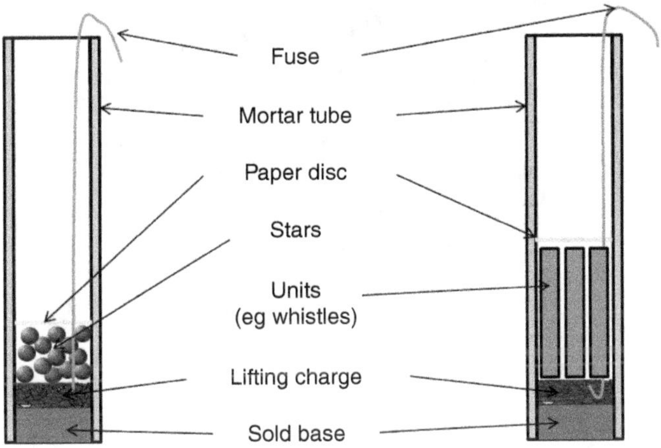

Figure 3.23 – Two types of mine in mortar

Subtypes of mines include:

***Bag mine, mortar mine** – A container with propellant charge and pyrotechnic units, designed to be placed in a mortar and to function as a mine. Container is typically a cloth or paper or plastic bag or cloth or paper cylinder.*

Report

Article containing pyrotechnic composition designed to produce a bang; it may also include a coloured delay element.

Subtypes of reports include:

***Flash banger** – A non-metallic case containing metal-based pyrotechnic compositions that may be used as pyrotechnic units in shells and other articles. The principal effect is a report and a flash of light.*

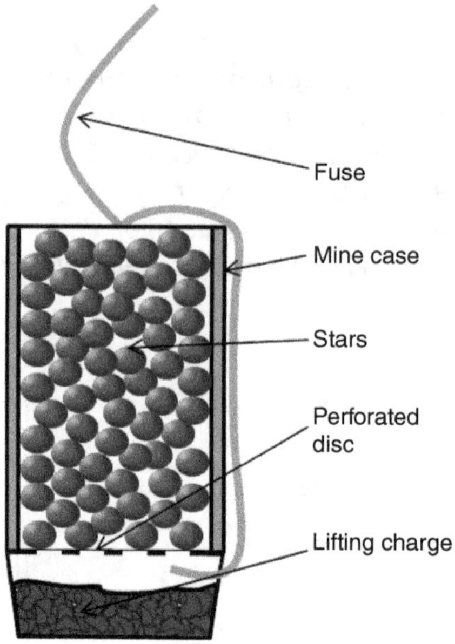

Figure 3.24 – Bag or "mortar mine"

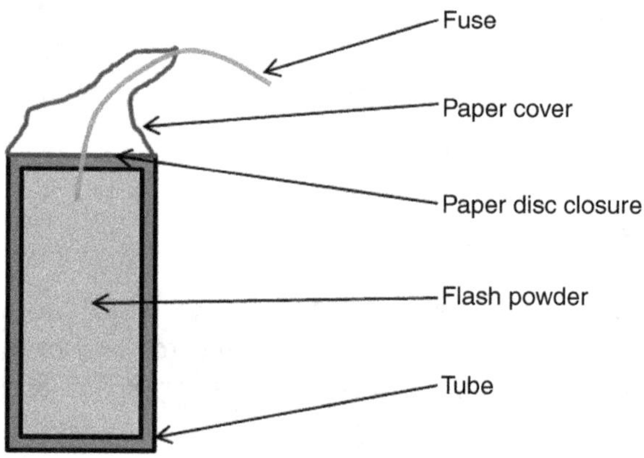

Figure 3.25 – Flash report or ground maroon

***Ground maroon** – A maroon without propellant charge and with or without delay fuse, designed to produce its report on the ground.*

***Maroon** – A firework containing pyrotechnic unit(s) or loose pyrotechnic composition and designed to produce a loud "bang" report as main effect. Not to be confused with bangers, as their design is similar to small shells or bombettes.*

Rocket

Article containing pyrotechnic composition and/or pyrotechnic units, equipped with a launching motor and stick(s) or other means for stabilization of flight, and designed to be propelled into the air. The principal effect is ascent, with or without additional visual and/or aural effects, and production of visual and/or aural effects in the air.

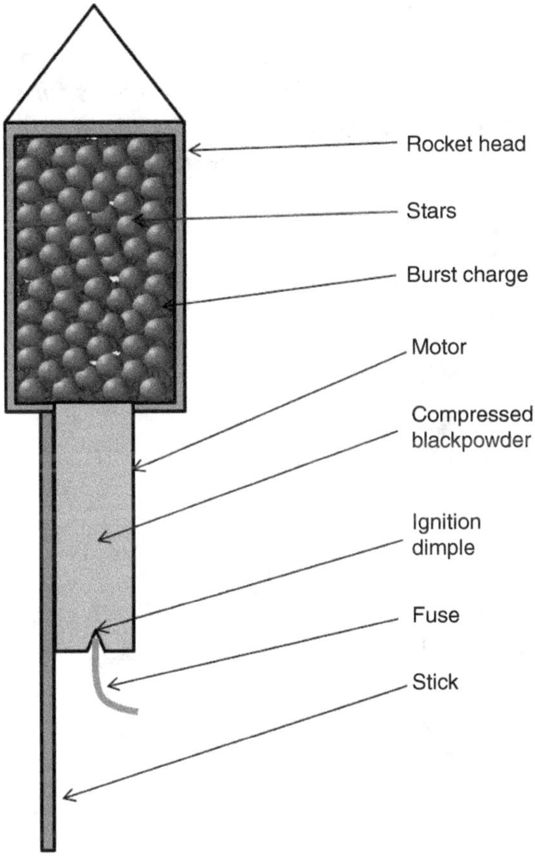

Figure 3.26 – Generalised rocket construction (with "German"-style motor)

The rocket motor may alternatively be formed on a spindle so that there is a "hole" along most of the length of the motor — this increases the surface burning area and thus the amount of gas produced — or it may be solid. A small dimple as shown in Figure 3.26 above is somewhat of a compromise — initial thrust is increased and the "dimple" extends somewhat as the motor burns.

Rocket motors for fireworks are usually made from blackpowder, or a blackpowder-like composition. Smaller rockets may be made from "whistle powder" — so that the rocket ascends and produces a whistle at the same time. There has been some move to use military-like propellants in firework rockets, the chief advantage being a dramatic decrease in smoke, but this practice is not yet widespread.

Subtypes of rockets include:

***Signal rocket** – A tube containing pyrotechnic composition and/or pyrotechnic units, equipped with a stick or other means for stabilization of flight, and designed to be propelled into the air to produce predominantly an aural effect. Signal rockets may also be coloured.*

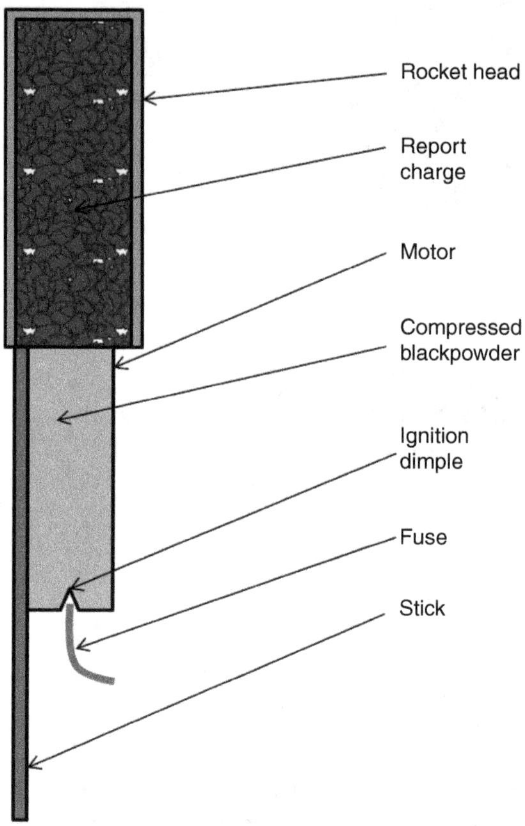

Figure 3.27 – Flash, signal or maroon rocket

***Flight rocket** – A collection of rockets designed to be propelled into the air from a frame or cone and ignited by a single or multiple fuses.*

Figure 3.28 – Flight rockets in Maltese display held in simple frame together with mortars supported by sandbags and metal frames. Photo: Nigel Village

Ignition of a single rocket causes all the other rockets to light due to the close proximity of the fuses. The ignition process accelerates (as in a chain reaction) and the rockets fly into the air, producing a large cone of effects.

The major safety issue with flight rockets is that the extent of the cone of flights is somewhat unpredictable, and indeed some rockets may fly at very low angles.

Parachute rocket – *An article containing pyrotechnic composition and/ or pyrotechnic units that contain subcomponents, some or all of which will descend on parachutes to the ground, and equipped with a launching motor and stick(s) or other means for stabilization of flight, and designed to be propelled into the air.*

Roman candle

A tube containing a single charge or alternate propellant charges, pyrotechnic units and transmitting fuses. The pyrotechnic units may be bombettes, comets, hummers, maroons, mini-mines, stars, whistles etc. The principal effect is ejection of the pyrotechnic units in succession, producing a series of visual and/or aural effects in the air.

There is a very large range of possible effects that can be deployed in a Roman candle. Some possible payloads are shown in Figure 3.30.

Roman candles for display use typically range in calibre from 18mm to 60mm.

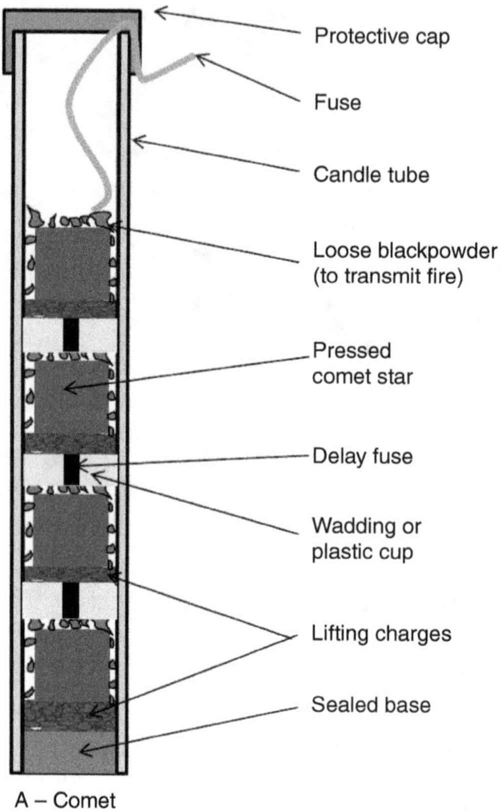

Figure 3.29 – Typical four-shot Roman candle with comet stars

Subtypes of Roman candles include:

***Shot tube** – A tube containing a single propellant charge and a pyrotechnic unit, with or without a bursting charge, with or without a transmitting fuse. The pyrotechnic unit may be a bombette, a comet, a hummer, a shell (including maroon shells), a whistle etc. (cf. mine).*

The shot tube, in various forms, also forms the basic component of multi-shot "cakes", but increasingly single-shot devices are used alone, especially when fired from structures. Significant development has been made to minimise the debris produced from such devices so that they also may be used indoors or in close proximity to audience or performers,

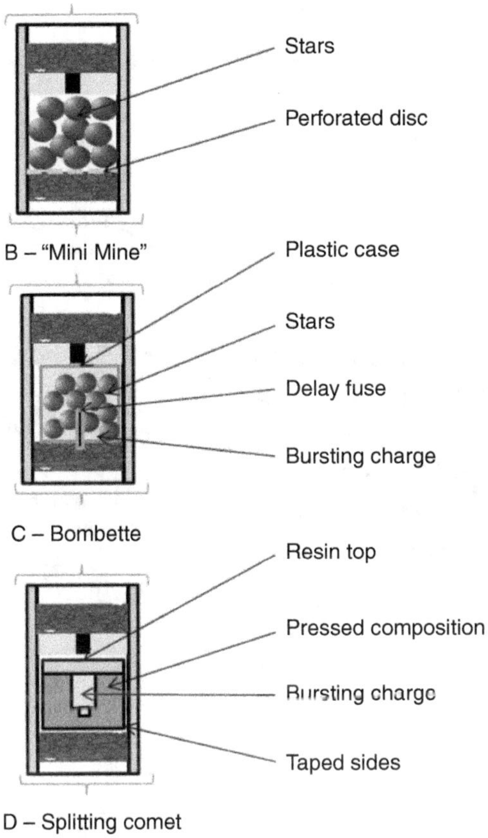

Figure 3.30 – Alternative payloads for a Roman candle; B — "mini-mine"; C — bombette; and D — splitting comet

in restricted areas or where damage to the structure could result from the normal debris produced.

Shell

A device with or without propellant charge, with one or more delays before bursting, pyrotechnic unit(s) or loose pyrotechnic composition and usually designed to be projected and burst at a distance from a mortar. The pyrotechnic units can be stars, butterflies, crackers, hummers, spinners/tourbillions, whistles etc. as well as report shells or other shells to produce multiple bursts simultaneously or sequentially.

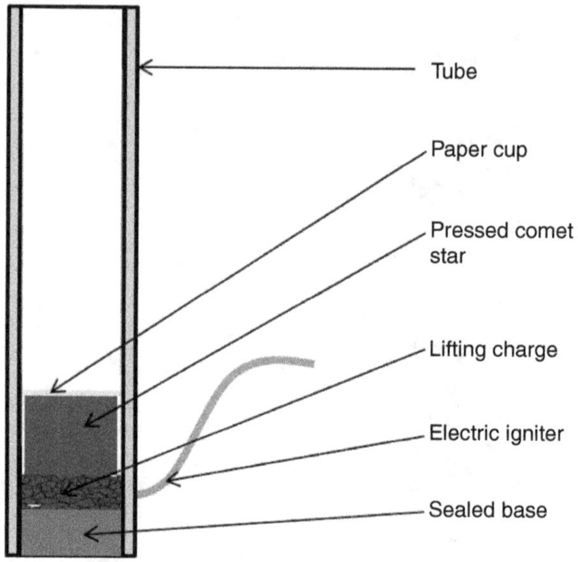

Figure 3.31 – Comet "shot tube" or "single shot"

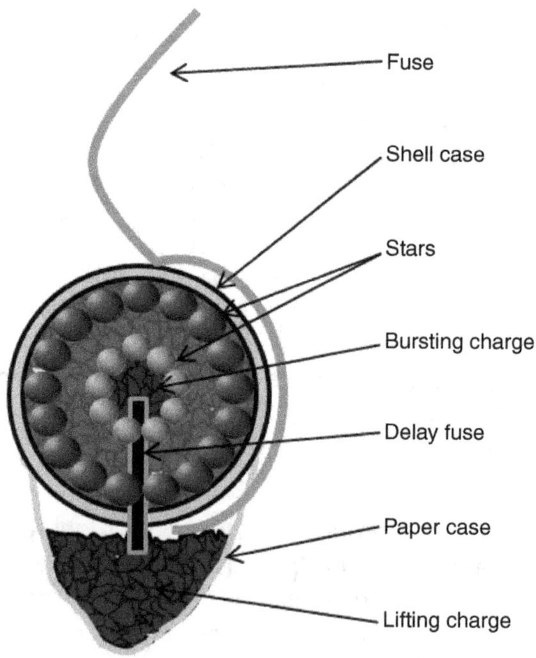

Figure 3.32 – Two-colour "oriental" style shell

Shells may be broadly spherical or cylindrical in design, and there are advocates for both types as well as practical considerations that favour one design over the other — for instance, it is easier to fit noise effects such as whistles into a cylinder than into a sphere (Figure 3.34).

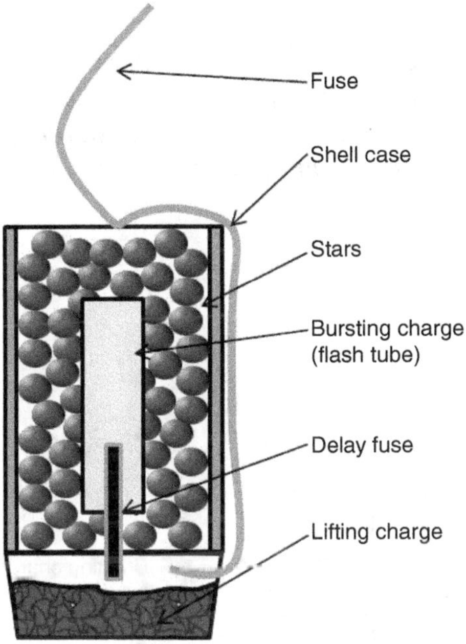

Figure 3.33 – Simple cylinder shell of typical European construction

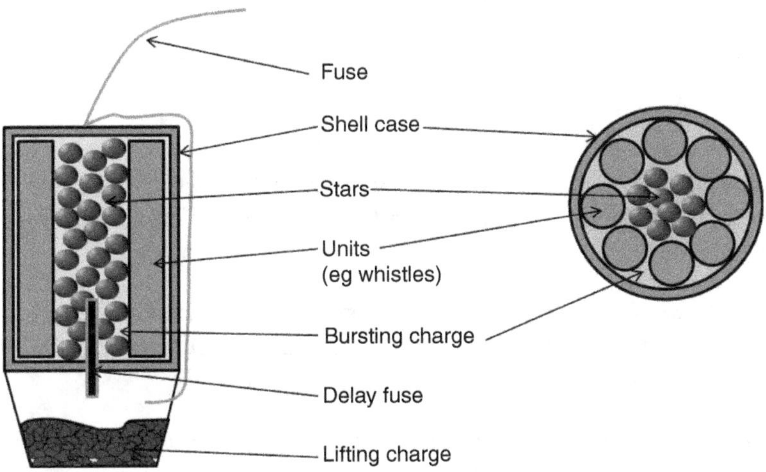

Figure 3.34 – Cylinder shell with stars and noise units

Subtypes of shells include:

Aqua shell, aquatic shell, nautical shell *– A spherical, cylindrical or other shell designed to be floated on water by means of a buoyancy device and fired from a mortar.*

Daylight shell *– A spherical, cylindrical or other shell designed to be fired from a mortar and containing sub-components that are visible in the daylight and/or components that produce an aural effect, emission of coloured light and/or smoke and/or aural effect.*

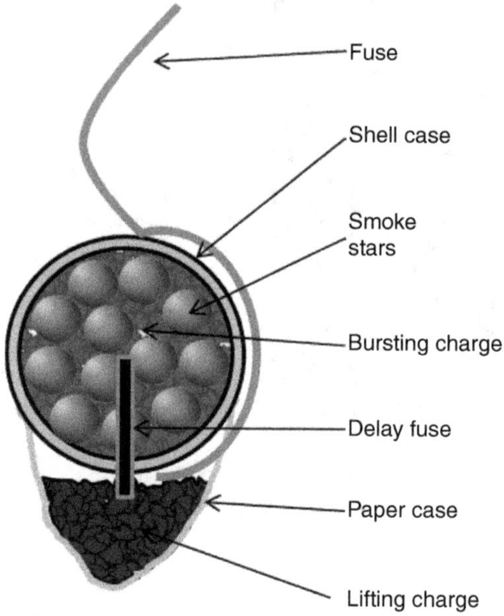

Figure 3.35 – Smoke shell

Maroon shell *– A maroon with or without propellant charge and with delay fuse, designed to be projected from a mortar and to produce its report in the sky.*

Parachute shell *– A spherical, cylindrical or other shell designed to be fired from a mortar and containing sub-components, some or all of which will descend on parachutes to the ground.*

Preloaded mortar, shell in mortar *– An assembly comprising a spherical, cylindrical or other member of the shell generic type, inside a mortar from which the shell is designed to be projected.*

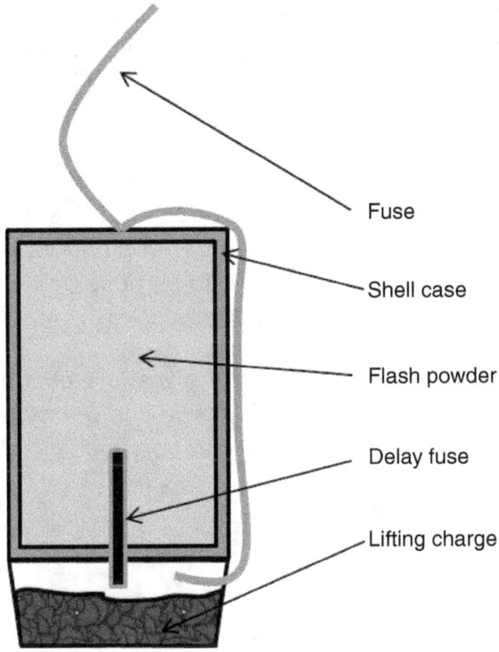

Figure 3.36 – Maroon shell or "salute"

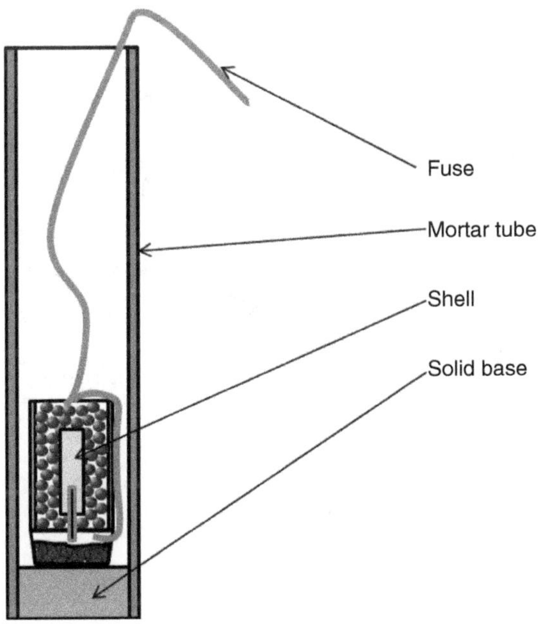

Figure 3.37 – Schematic of shell in mortar

Small-calibre shells in mortar are effectively single-shot bombette candles — the distinction is generally taken as to whether the payload has an integrated lifting charge (i.e., a shell) or the lifting charge is separate (as in a candle).

Multi-break shell *– A shell with several discrete elements and with or without propellant charge, with delay fuse and bursting charge, pyrotechnic unit(s) or loose pyrotechnic composition and designed to be projected from a mortar and to function sequentially or simultaneously by the lighting of multiple internal delay fuses.*

Repeater shell *– A device with several discrete elements and with or without propellant charge, with delay fuse and bursting charge, pyrotechnic unit(s) or loose pyrotechnic composition and designed to be*

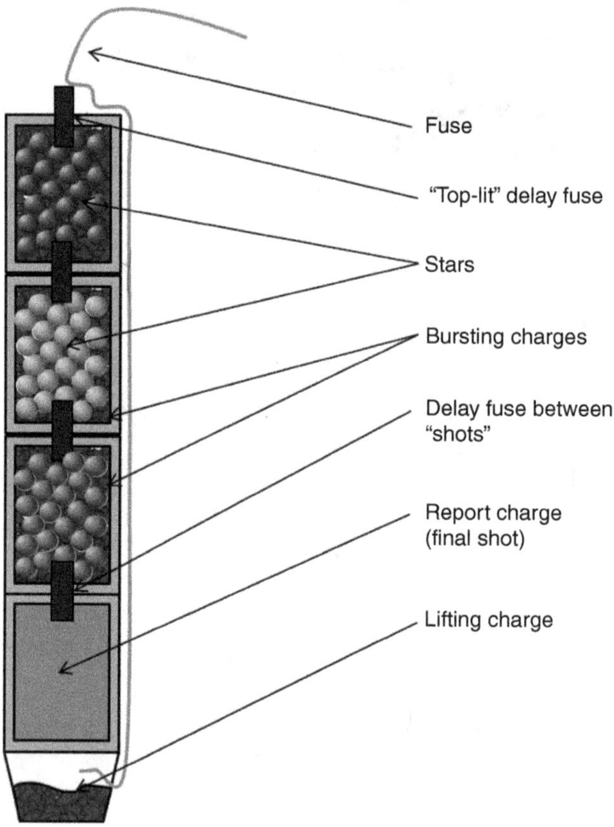

Figure 3.38 – Four-break repeater shell (colour/colour/colour/maroon)

projected from a mortar and to function sequentially by the lighting of the internal delay fuses by the functioning (burst) of the previous device.

Multi-break and repeater shells are often "top-lit" — where the initial fuse ignites the first internal delay fuse for the first break and also ignites the lifting charge. This is in contrast to a simple shell where the initial fuse ignites the lifting charge, which in effect ignites the internal delay fuse.

There is an inherent danger with the "top-lit" approach, that is, if the top fuse is ignited and for some reason (damp fuse, break in fuse etc.) the lifting charge is not lit and thus the shell is not propelled into the air and the burst occurs when the shell remains in the mortar, then there is a risk of bursting the mortar.

In addition, the desire by some to produce larger and larger multi-break shells with more and more "breaks" requires the shells to be fired with larger and larger lifting charges (and hence the mortars must be constructed to contain the forces of the lifting charge). There is a danger that the final "shots" could burst as the shell has passed the apex of its flight or indeed, as has been seen at several demonstration events, when the shell has returned to ground level. Needless to say this is undesirable!

Complex shell – *A shell composed of several discrete elements designed to be projected from a mortar with a single lifting charge and to function sequentially or simultaneously. The subtypes of complex shells include the following articles:*

Peanut shell – *Device with two or more spherical aerial shells in a common wrapper propelled by the same propellant charge with separate internal delay fuses.*

Shell of shells (spherical) – *Device with or without propellant charge, with delay fuse and bursting charge, containing report or other shells as subcomponents and designed to be projected from a mortar.*

Smoke/fog generator

An article containing smoke-producing pyrotechnic composition or heat/gas-generating composition designed to evaporate a substance or disperse hygroscopic particles and designed to function on the ground or fixed to a support. The casing of the article can be made of different materials. The principal effect is emission of white or coloured smoke/fog without any aural effect.

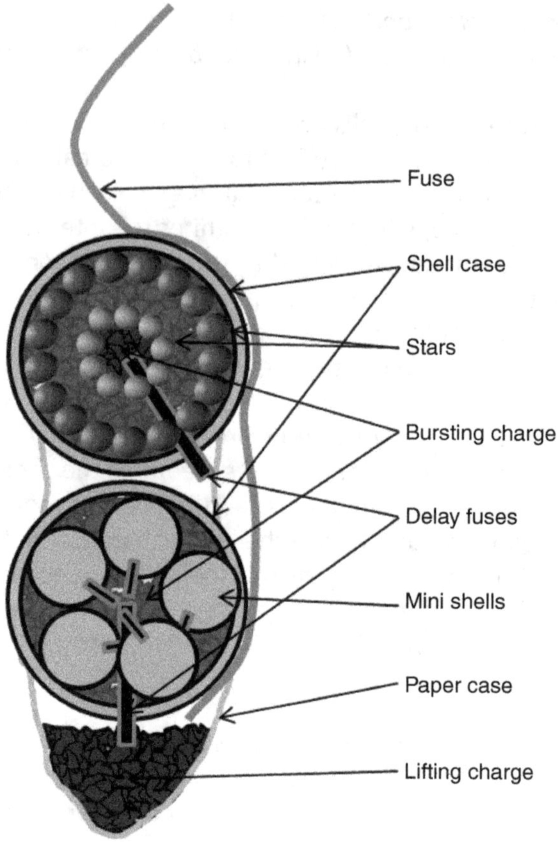

Figure 3.39 – Peanut shell (two-colour and shell of shells)

Smoke generators are not used extensively in outdoor firework displays, although they may be used in conjunction with the use of lasers to supplement the display.

Other firework types and effects

There are other types of fireworks that may be used in outdoor firework displays, but their usage is rare and specialised and not considered further here.

Within each firework type there are a large number of possible variations of size, colour and effect. A cursory analysis of a single Chinese manufacturer of shells revealed that within the range of sizes offered (2" to 16") there were no less than 62 different generic subtypes of

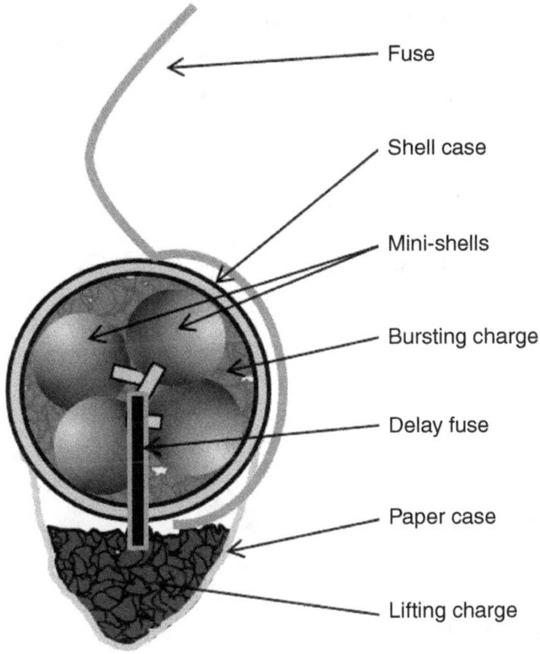

Figure 3.40 – Shell of shells

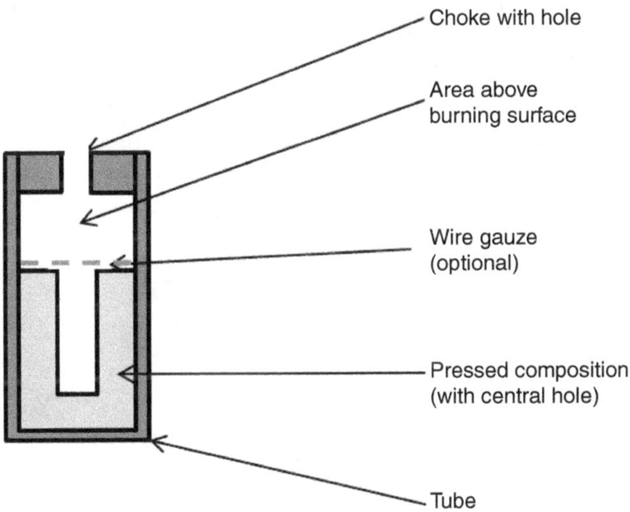

Figure 3.41 – Smoke generator

shells (for instance single-colour, double-colour, colour-change etc.) and over 500 different colour combinations or effects with "green" in their description, including familiar types such as:

- Green peony
- Green chrysanthemum
- Green and silver chrysanthemum
- Green to silver chrysanthemum
- Green chrysanthemum with silver pistil
- etc.

And more esoteric types such as:

- Green rings
- Green butterflies
- Green octopus
- Green bow ties
- Green palm with tail
- etc.

For shells with only green effects this meant over 3500 different shells from this single supplier. Across all of the possible colours, subtypes and size variations of a single firework type, the number of combinations can easily reach many tens of thousands. The situation for multi-shot items ("cakes") is potentially even worse.

This produces problems not only for the manufacturer, but also for the display designer and, eventually, the enforcement authorities. Although the variety may be academically interesting it is doubtful that a viewer would recognise the difference between closely related types. Furthermore, even from one manufacturer where a range of colours is available — say the gradation of

- Yellow
- Lemon
- Citrine
- Lime
- Green
- Emerald

then the number of potential types increases vastly!

Only by clever use of the colours, and deliberately emphasising the change from one colour to the next, will the average viewer realise there is a difference at all!

Firework effects that cause potential problems

There are a number of generic effects that potentially cause the greatest problems in designing and firing a display. This is not to suggest in any way that such effects should be restricted in manufacture or in use, but merely that extra care should be taken when using them. In almost every case the potential problems arise because of the long-burning nature of the effect, which can allow the burning pyrotechnic composition to reach ground level. This can either be in "normal" functioning, and even by design, or particularly if the firework does not function correctly (e.g., does not reach the design height), and stars that were meant to be extinguished by the time they reach the ground are not.

Table 3.2 – Potential problematic effects

Effect	Most usually found in	Issues
Kamuro Brocade Crown Chrysanthemum etc.	Shells Rockets (Mines)	Stars falling to the ground. Drifting of effect.
Strobe	Shells Rockets	Stars falling to the ground. Drifting of effect.
Glitter	Shells Rockets Roman candles	Some of the yellow glitter varieties have particularly long-burning drossy fallout.
Parachute effects	Shells Rockets	If parachute fails to open — then stars may burn on the ground for an extended period. If parachute opens, then the star may drift significant distances downwind.
Falling leaves	Shells	Similar but less extreme than parachute items.

The effects of general "fallout" will be considered further in Chapter 7.

The display designer needs to take account of all the possible variations of types and all the possible variations of effect when designing the show. The display firers also must be knowledgeable of the effects so that they can adapt the display, or propose cancellation or curtailment at the display site on and before the time of firing. Clearly this is a daunting task, and it is better to plan the display in such a way and with objective modification or cancellation criteria so that a subjective decision is not left to the firers (who may not be the display designers). Much of the remainder of this book addresses the determination of such objective criteria.

Notes

1. EU Standard for fireworks — See for a summary. http://europa.eu/legislation_summaries/consumers/consumer_safety/l11024a_en.htm
2. M Davies, Journal of Pyrotechnics, 21, 2005, 1–12 — See also http://www.jpyro.com/wp/?p=120
3. M Davies, Journal of Pyrotechnics, 27, 2008, 42–29 — See also http://www.jpyro.com/wp/?p=577
4. See for example "Glitter Chemistry," C Jennings-White, Pyrotechnic Chemistry, Ed. K L and B J Kosanke, Pub: *Journal of Pyrotechnics*, ISBN 1-889526-15-0 — See also www.jpyro.com/wp?p=291 and Lloyd Scott Oglesby – *Glitter, Chemistry & Technique*, Pub: American Fireworks, 1989, ISBN 0929931017.
5. Strobing reactions in solution — See for example Doris Kolb, *J. Chem. Educ.*, 1988, 65 (11), p 1004.

Chapter 4 – Display types

Tom Smith – Davas Ltd

The range of sites from which a display has to be fired and viewed is almost as diverse as the range of fireworks available to be fired. In this chapter we will examine the opportunities and constraints each type of site poses, and subsequent chapters will then highlight the consequent process to determine the types of fireworks to be used, the site and product-specific analysis of risk at that site with the chosen fireworks (the two processes necessarily run in parallel) and the contingency measures that need to be adopted for modification, curtailment or cancellation of the displays when they are fired.

We have to accept that the impact of fireworks is often related to the relative proximity of the audience. It may be safer to fire a display a very long way away from any potential hazard (people or buildings), but it is rarely practical or aesthetically pleasing to do so. Even for the largest-scale events where the overall intricacy of the design may only be appreciated by those at a significant distance (or, more likely by those watching remotely via television or the Internet), it is likely that there will still be people local to the display firing site who, as well as having a completely different appreciation of the event, are exposed to risks from the event.

In general, the larger the event, the less flexibility there is in the choice of firing site although conversely perhaps, the greater the potential for modification of the display at firing time. In general also, the largest displays actually have the lowest proportion of the budget in the actual fireworks, there being so many other demands of staging such events that are only of limited interest here (crowd management, emergency planning, TV production etc.).

The smaller displays tend, in contrast, to have some flexibility of firing position and little flexibility of the actual fireworks that may be used. The latter may be due to factors such as:

- The types of fireworks actually available to the sort of people firing the display (e.g., non-professionals)
- The need to purchase the fireworks prior to the display and therefore the lack of alternatives available (which might be available to the professional at short notice)

Figure 4.1 – Brock's poster from c. 1900

Irrespective of the nature or scale of the event the same processes should be adopted to determine the suitability of the display site and firing site, and the fireworks to be used.

1. Survey the site(s), paying attention to:

- Possible alternative firing positions (if any)
- Flexibility on the chosen site(s) to adapt to conditions prevailing at firing time

- The desired (but not always fixed) position(s) of the audience
- Other local hazards
- The prevailing meteorological conditions (which may be quite local)
- The physical nature of the firing site (e.g., grass or concrete, level or sloping etc.)
- Constraints the site imposes (e.g., firing site above or below audience)
- Opportunities the site offers for creative use of fireworks

2. *Then determine:*

- The most likely and various other (including worst-case) scenarios for fallout
- The distances to audience(s) and other hazards

3. *Now use the information to determine the fireworks types and calibres that can be used:*

- Under normal and foreseeable meteorological conditions at the chosen firing site
- Under a variety of other conditions

4. *And then (and only then) plan the overall display, taking into account:*

- Organiser's/producer's requirements (if applicable)
- Your own artistic input (and from others)
- Within the constraints of types/calibres determined from the process above
- With a view to being required to fire a curtailed display if the conditions dictate
- Within budget!

What this usually means is that a conservative display must be designed so that it may be fired under the greatest possible range of conditions OR that the firing position and/or the types of fireworks actually fired must be flexible to allow last-minute changes. In practice such flexibility is rare and hence the design of the displays has to reflect a large number of possible conditions in which they will be fired. This is not to say that contingencies should not be incorporated into the display design. Whether a display is fired manually or electrically it must be possible to abandon the firing of particular fireworks by type, calibre or effect. For instance, it may be necessary to remove larger-calibre shells (because the "normal" fallout would extend too far), or fireworks with long-burning stars (e.g., Kamuro).

Typical sites and display scenarios

The following section looks at eight typical display scenarios, ranging from a small amateur or professional display, to a complex display fired from many points simultaneously. In each case potential opportunities and constraints are identified. Obviously, these are just illustrations and every display site is different — and even for two displays on the same site a large number of other factors will mean, ultimately, that two shows are rarely the same!

Summary of the scenarios

A. Display for school, fired from school playing field at the side of school, audience in front of school building
B. Display for traditional UK November 5th celebrations — including a bonfire
C. Display for music festival, firing site on island, audience on shore of lake
D. Display fired from dockside (concrete) watched across the water

Table 4.1 – Display for school, fired from school playing field, audience in front of school building

	Issues
Firing site	On school playing field — nothing must be "dug in" to the playing surface
Audience position and number	Approximately 200 people of various ages (parents, children — some very young, grandparents) in front of school buildings and well marshalled
Firers type	Amateur or small professional display, usually hand fired
Fireworks types	Large consumer or small professional items
Things to look out for	Adequate barriers between crowd and firing area (and kids crossing the barriers) Limited firing area Properties surrounding the school area
Possible opportunities	There is a significant "width" available — perhaps the drama can be achieved by using this "width" instead of excessive noise.
Possible Constraints	Noise levels (small children and local neighbourhood)

E. Pyromusical display for large outdoor concert, firing site behind stage
F. Display fired from rooftop of city centre multi-storey car park watched from ground level around the base of the buildings
G. Display fired from three barges on large tidal river
H. Display fired from ground level and from adjacent sports stadium roof for audience inside stadium and TV audience

This display is typical of many small amateur displays using consumer fireworks. The primary audience is small children, so it can be useful to start the display slowly and with few noisy effects. As the display progresses, so can the intensity and impact of the display. The range of fireworks used is relatively restricted and often comprises the firing of a succession of multi-shot devices (cakes) punctuated by the use of rockets (for emphasis) or set-pieces (to provide a calming, low-intensity, and relatively cheap interlude). The finale can be quite "wide" with simultaneous firing of, perhaps, two or three identical cakes across the display site.

As always, the site should be decided well before the event, and the demands of firing a display at all incorporated into the overall event plan. Too often we have seen the fireworks forced into an inappropriate area because, for instance, the catering stands were perceived as fund earners and therefore given the most prominent position.

The firers (especially if they are amateur firers) should familiarise themselves with all of the fireworks prior to the display, in daylight and in safe surroundings, paying particular attention to the instructions and any requirements for equipment, and to the position of the fuses. A firing order should be decided and jobs allocated. If the display is amateur fired there can be a temptation to involve too many enthusiastic people. As a general rule two or three people would be quite sufficient to actually fire the display.

This display is illustrated further in Chapter 19.

Table 4.2 – Display for traditional UK November 5th celebrations — including bonfire

	Issues
Firing site	On the edge of a large playing field
Audience position and number	Typically several thousand people, predominantly on one side of the firing area. Occasionally the audience area might extend around the display site — this is not desirable.

	Issues
Firers type	Amateur or small professional display. Hand firing and some simple electrical firing may be used.
Fireworks types	Large consumer or professional items
Things to look out for	Adequate barriers between crowd and firing area (and kids crossing the barriers) Limited firing area and built-in inflexibility of the site
Possible opportunities	Lancework and set-pieces may often be incorporated, positioned nearer the audience than the main display.
Possible Constraints	Limited firing area Little flexibility in case of change of wind direction

This is typical of UK November 5th displays, where there will be a variety of ages present. The sites are often well established and offer little flexibility. If there is a bonfire, then all possible efforts should be made to have the bonfire lit after the fireworks (although this is not traditional) because the light from the bonfire will ruin the impact of the colours of the fireworks. In addition, if the bonfire is positioned within the extended firing area, care must be taken to ensure it does not ignite prematurely from stray firework sparks, and conversely, if the bonfire is lit first, that sparks from it do not fall within the firework firing area.

The range of fireworks used at such displays is normally quite extensive and this adds variety to the display — which can often be quite long. Rarely is music played during the firework display itself and hence the need for synchronisation is absent. Consequently a typical amateur or professional display will be formed from a number of sequences where one type of fireworks is fired at a time, building in intensity with larger and more extensive aerial effects; then the sequences are repeated. As in all displays a good finale is essential, and very often such traditional displays are started and finished with maroons.

As with the previous display it is imperative that the firers familiarise themselves with the firework types, their effects and if possible their duration before the display, and that firing roles and a firing order are established so that there is minimum possibility of problems arising during rigging or firing. It is important that the display "flows" and that gaps are kept to an absolute minimum. We often recommend that for hand-fired displays one person is established with a firing frame loaded with medium- or large-calibre rockets. If, for whatever reason, a gap happens, then this person can fire the rockets relatively slowly until the normal firing sequence is resumed. Any rockets remaining unfired at the commencement of the finale can then be fired to boost the finale even more.

Table 4.3 – Display for music festival, firing site on island, audience on shore of lake

	Issues
Firing site	On shore of island
Audience position and numbers	Typically 1000–2000 people on banks of lake facing the island
Firers type	Medium-sized professional display. Hand firing and electrical firing may be used.
Fireworks types	Large professional fireworks with some special effects for use on the water
Things to look out for	If any equipment is "dug in", then the water table can be reached and the equipment will become wet.
Possible opportunities	Reflections of the fireworks in the water. Low-level effects will be reflected very well, but the higher-level fireworks may not be reflected at all.
Possible Constraints	The firing site is fixed — so if the wind is towards the audience the display may have to be curtailed. If the display is fired over many years, care should be taken to ensure the firing area is not encroached upon by trees and other vegetation.

Some of these displays will incorporate novelty devices (such as relatively complex and animated lancework) to add both variety and changes in tempo, and as a result in a fixed-length display, to allow a greater proportion of the overall budget to be fired as a finale. A display of this type is illustrated further in Chapter 19.

Although this display is accompanied by but not synchronised to music, an appreciation of the music is essential. It would be totally inappropriate to be firing loud items during quiet sections of the music! Where the music is familiar to the audience there is an expectation that the general tempo and intensity of the music will be mirrored by the tempo and intensity of the fireworks, and indeed if this is successfully achieved many in the audience will believe there is more synchronisation than there actually was!

Where the music is pre-recorded, such apparent though low-level synchronisation is relatively easy to achieve. Where the music is live this is more difficult to achieve — timings from performers and conductors are notoriously inaccurate, and they wish to maintain their artistic flexibility and react to the audience as appropriate.

However, it is important that certain long-duration items (such as cakes or Roman candles) are fired at such a point that they do not extend into an inappropriate section of the music — for instance if they are fired for the last part of a loud and boisterous section that is to be followed by a quiet and subtle movement of music, then they must be fired at a point no nearer to the transition than their firing duration. This seems obvious in theory but often is overlooked in practice. Instead, the final few bars of the loud section should be accompanied

Table 4.4 – Display fired from dockside (concrete)

	Issues
Firing site	Dockside in front of three-storey buildings in part of an old dock
Audience position and numbers	Typically 10,000 to 15,000 people across the dock and at other locations — reflections in the water
Firers type	Medium-sized professional display. Hand firing and electric firing may be used.
Fireworks types	Large professional fireworks with some special effects for use on the water
Things to look out for	Water users (boats, canoes etc.) encroaching into the safety area Debris falling onto the buildings
Possible opportunities	Often it is possible to fire from several points simultaneously. It is also possible to fire "outwards" from the dock with Roman candles, waterfalls etc. — provided they are projected into a safe area.
Possible Constraints	Firing from solid concrete surface — fireworks will need to be self-supporting, or racked and supported by, for instance, sandbags.

by short-duration fireworks (shells, rockets, mines etc.) so that firing may be ceased immediately.

Again, prevention of gaps is critically important, and by careful design some of the display can be reserved for this task whether the display is manually or electrically fired.

The major issue with this display is the requirement to fire from a solid surface and hence the need for all firing equipment to be self-supporting. A variety of methods are available for this and are discussed in Chapter 8. If a firework were to fall over, then it may impact on the buildings behind the firing site, or be projected over the water towards the audience.

The audience is often on flat ground on the other side of the dock, and the possibility for seeing the water surface (and hence reflections) may be limited to only the first row of the audience. Distances across water are often deceptive, and care must be taken to ensure that the "safety" distances are not compromised.

In addition, in surroundings such as this there is often a great disparity between the wind strength and direction at water level and at the height at which, for instance, shells burst — and hence the fallout pattern can be quite distorted and extended.

Access control to these areas seems to be a particular issue, the public being resentful that their normal walk is being disrupted! It can be

Figure 4.2 – Setup on solid surface — Plymouth Mountbatten Breakwater

extremely useful to have notices to inform the public of what is going on, to reassure them that it is only a temporary disruption to their lives, to give information when the display is to be fired and where the designated viewing areas, and to reassure them that you have the necessary "matches" to fire the display!

This display is a synchronised firing to live music and requires considerably more attention to the order of firing than the previous displays. If possible, accurate timings should be obtained from the orchestra or band to enable the preliminary planning of the display. Depending on the budget of the display there may be opportunity to attend rehearsals

Table 4.5 — Medium-scale pyromusical display for large outdoor concert, firing site behind stage

	Issues
Firing site	Grassed area behind stage Some flexibility to move the firing site dependent on the weather conditions
Audience position and numbers	Typically 5,000 to 8,000 people in front of stage
Firers type	Medium-sized professional display Predominantly electric firing
Fireworks types	Large professional fireworks
Things to look out for	The display will not appear directly over the stage except for a very small proportion of the audience.
Possible opportunities	Use of the stage for some close-proximity effects
Possible Constraints	Site position is fixed so that the fireworks appear above the stage for the audience. The height of the stage restricts the lowest height of the fireworks.

and make accurate timings, although experience shows that these may not be followed exactly on the night!

If the stage is some distance from the firing site, it is essential to have a loudspeaker "monitor" led to the firing site so that any delay from the speed of sound is minimised.

A really good appreciation of the music is essential, and the tracks, movements or pieces to be used should be listened to by all of the firers prior to the display and notes taken to ensure that the various sections (tempos, intensity etc.) are identified.

Usually fireworks are only used at the end of the concert, not only because this is the natural conclusion and highlight of the event, but because only at the end of the display will the light levels be such that fireworks can be used successfully. Very often it will be the intention of the performers to include an encore — in which case this should be planned before the display, as otherwise it can look distinctly anti-climactic! If the budget is tight, perhaps only a few seconds of fireworks are needed at the very end of the encore to play psychological tricks with the audience. If the fireworks are more intense than they were expecting, they are likely to be more impressed than by a low-intensity, dragged-out display during the encore.

Table 4.6 — Display fired from rooftop of city centre multi-storey car park

	Issues
Firing site	Multi-storey car park
Audience position and numbers	Typically 15,000 to 20,000 people around the base of the car park and all around the city centre
Firers type	Professional firers
Fireworks types	Low-debris materials — If possible self-consuming
Things to look out for	"What goes up must come down" — there will be minor debris from all fireworks that can travel significant distances
Possible opportunities	If more than one site is available the lack of aerial material can be compensated for by simultaneous firing from more than one venue.
Possible Constraints	The actual firing area may be quite restricted. The wind can funnel between and around buildings and may lead to extended and unpredictable fallout areas.

It is my belief that this sort of display works best when the music is familiar — so that if the audience expects, for instance, an increase in tempo and intensity, the fireworks should accordingly increase in tempo and intensity. If they are straining to appreciate a new or unfamiliar piece of music and are watching fireworks that do not appear synchronised at the same time, the experience will be less satisfying.

High levels of synchronisation are possible with enough planning and a big enough budget — and the ways of achieving this are discussed further in Chapter 18.

The critical feature of this display is the fallout from the fireworks fired and where it will land. Of course the firing from a rooftop also means that all firing equipment will need to be self-supporting, as in Display D above.

Whatever types of fireworks are used there will be some fallout. There are a few stage effects in which there is literally no debris produced, but it is likely that a major firework display will require an extended range of fireworks to be used. It is generally not appropriate to use shells in this situation — although the failure rate of shells is low, the possible consequences of a dud shell falling to the ground are very great and thus the risk is unacceptable.

Even traditionally low-debris fireworks such as Roman candles do produce a significant amount of inert debris — the wadding between "shots" can cause significant injuries if it falls on an area closely packed

Table 4.7 – Display fired from three barges on large tidal river

	Issues
Firing site	Three barges moored or held by tugs
Audience position and numbers	Typically 40,000 to 60,000 people spread along the river banks.
Firers type	Electric firing with synchronisation between barges.
Fireworks types	Typically shells, mines and Roman candles.
Things to look out for	The wind often "funnels" along the river at water level, but can be quite different in strength and direction at higher levels above building height.
Possible opportunities	With three barges the display can be varied by firing simultaneously and symmetrically from the barges, or deliberately choosing to introduce asymmetry into the display. The central barge can contain the major aerial materials, as it is likely to be the central barge that is farthest from the audience.
Possible Constraints	It is difficult to position the barges accurately in the tideway, and they may swing as the tide changes direction. Also, the state of the tide (high or low water) at the time of firing may affect the visual impact of the display.

with audience. The use of "single-shot" items, where there is no such wadding between the shots, is of course desirable but may neither be practical nor economical. Even using such material there is likely to be some minor debris (paper cups holding the stars in place within the firing tube, pieces of the fusing system etc.) that will fall to the ground.

Given the elevated firing position and the way the wind moves around the buildings (especially if the firing site is in a mass of similar-scale buildings), the fallout area can be extensive and unpredictable. It is usually unrealistic to try and remove people from the ground around the base of the buildings, and the choice of fireworks to be fired can thus be very restricted.

This display requires synchronisation of the firing from the three barges — but this does not necessarily mean that the displays on each of the barges will be identical. The use of the three firing positions, often quite well separated, can allow the display designer some creative opportunities that are rarely available on land-based sites. Possibilities are further discussed in Chapter 11.

The three barges could be linked together physically by the firing system, but it is more usual to fire each barge autonomously with the firing systems synchronised by broadcast time code or use of GPS locking (see Chapter 18).

Table 4.8 – Display fired from ground level and from adjacent sports stadium roof for audience inside stadium and TV audience

	Issues
Firing site	Multiple sites
Audience position and numbers	Extensive — both within and outside the stadium itself. Typically 20,000 to 80,000 within the stadium and a further 50,000 outside the stadium. TV audience in millions
Firers type	Electric firing synchronised between sites
Fireworks types	A full range of "off-the-shelf" and bespoke items are available.
Things to look out for	It should be possible to curtail the display at any of the sites independently without affecting the overall display unduly. It should be possible to cease firing particular types/calibres of fireworks as the conditions dictate. It may well be that the artistic demands of the event producer may not require a display in the traditional sense. Increasingly, fireworks are used to punctuate particular moments in the event and there may only be extensive use of fireworks in the finale segments.
Possible opportunities	With an extensive firing area there are possibilities for deliberate asymmetrical firing — for instance "chases" across or between structures or firing points. However, this necessitates close cooperation and prior planning with, especially, the broadcasters to ensure that these are exploited and appreciated to their full extent.
Possible Constraints	People around buildings The larger the firing area, the larger the amount of space where the audience cannot be The flexibility of firing positions is reduced to zero — it may be that whole firing sites will have to be curtailed during the display if the conditions dictate.

This is the most complex example and requires considerably more planning to meet the requirements of both local audience and broadcast media — the demands of the two are not always the same!

For this scale of event the planning will have started at a very early stage, and the demands of the event producers will largely dictate the types and extent of the display produced. Furthermore, the ideas of the event producer may not even be feasible or safe, but safety must be a paramount concern as the negative publicity and potentially serious consequences of any incident outweigh the normal desire of the designer to use larger-calibre materials.

Most such events do not rely on live music being broadcast, but instead use pre-recorded music and effects, to which a live orchestra or band may mime. Although this may be artistically dishonest, it is critical to have sub-second timings available and adhered to. This precision should also extend, where possible, to movements of people and to speeches — although this is rarely achieved. Instead, a series of "milestones" are developed and the firing system must be flexible enough to react to changes in these milestones as the event progresses.

Liaison with the broadcast media is essential, and the firework display designer should help them decide the ideal image angles and extents (wide angle, close-ups etc.) to maximise the visual effect to the broadcast audience. Designing a display for the media often contradicts the needs of designing for the local audience. Compromises will have to be made, but often the demands of the broadcast media predominate.

Computer-generated imaging (CGI) or video can often be used to illustrate the display and to determine the best filming positions and angles, but it must be appreciated by all that the real-life firing and the CGI will rarely be exactly the same. The highly publicised cutting between live and CGI effects at the opening ceremony of the Beijing Olympics in 2008 have, to an extent, given the use of CGI planning a bad name, but it would be unrealistic to expect an ad hoc approach to a display of this size or importance, and the use of CGI planning will inevitably increase.

Examples of very large, complex displays such as this are given in Chapter 20.

The next steps

Once the possibilities and constraints of a display site and the display requirements have been investigated, then the display can be designed. However, the safety of the display is still not assured. In order to reduce the possible risks to acceptable levels we need to consider a variety of other factors — and these will be dealt with in the following chapter.

Chapter 5 – Safety issues and risk assessment

Tom Smith – Davas Ltd

This chapter, much of which has been developed from previous publications by the author, addresses the fundamental safety issues for operators, the audience, the area from which the display is to be fired and anything else that may be affected by the rigging and firing of the display. This chapter contains summaries of papers published by the author and others elsewhere. Further details should be sought from the full papers.

General issues

Fireworks are explosives: they have the potential to cause harm to operators, the public and to structures. In our modern litigious society it would be unreasonable not to take safety seriously, but the level of planning and extent of precautions developed are inevitably dependent on the scale of the display, the budget for the display (and for the event as a whole) and the nature of the display site.

Furthermore, it is imperative that the safety implications of many aspects of the display are considered alongside each other — it is not acceptable to consider any aspect in isolation. Hence the potential impact of all of the following is important:

- The safety of the firework display operators
- The safety of the audience
- The safety of bystanders who may be impacted by the display — but are not the "target" audience
- Any impact on local hazards
- The safety of structures used for firing the display or otherwise affected by the display
- Adverse publicity for the company and the event organisers/venue in case of incident
- Economic consequences of an incident

As a consequence of all the levels of involvement it is vital that all parties involved in a display "sign up" to the management and control of safety for the event. This applies equally within the display company itself — it is not acceptable for the company management to pass all responsibility for maintaining safety to the workers. If good management systems and practices are in place, and the workers understand and adhere to these practices, then safety can only be enhanced. Too often we witness people all too ready to take credit for an outstanding event who would not, if the event did not work as intended, be prepared to take an equal level of responsibility.

This is the last time the word "safety" will be used in this context — any normal event using fireworks is never "safe"; all we can hope to achieve is to reduce the risks to an acceptable level.

Risk assessment

Risk assessment is a much-abused concept. Assessing the risks is not the same as "doing a risk assessment". The latter term has become devalued and in many cases it simply involves photocopying the last risk assessment! Assessing the risks is a serious task, and although in any operation, for instance a firework display, many factors remain constant, there are always site-specific factors that must be addressed.

Constant factors may include:

- The range of fireworks used
- The methods of erecting mortars
- The firing system

Factors that change from site to site, and crucially from event to event, include:

- Local weather conditions
- The physical site, for instance: Can mortars be dug in, can angle irons be used, or does everything have to be supported by sandbags?
- Constraints of the site, for example: where there is plenty of room for varying the firing position, the choice of fireworks may be made knowing that the site can be adapted with knowledge of likely wind conditions during the display — for instance, barges held by tugs may be moved to maximize the fallout area. On the other hand, where the site is fixed, the choice of fireworks may be conservative and dictated by the "worst-case" scenario.
- Local hazards (e.g., gas cylinders in the fallout zone)

That is not to say that previous risk assessments are not valuable. Over time, previous risk assessments form a valuable resource, especially where they have been shown — as a result of a "near miss" or real incident — to be lacking. Revision and modification of existing risk assessments in the light of extended experience are probably the most valuable revisions possible.

Assessing the risks does not stop when a risk assessment is written. The process is iterative and risks are not adequately controlled if the process is stopped at any point. Old, out-of-date risk assessments are

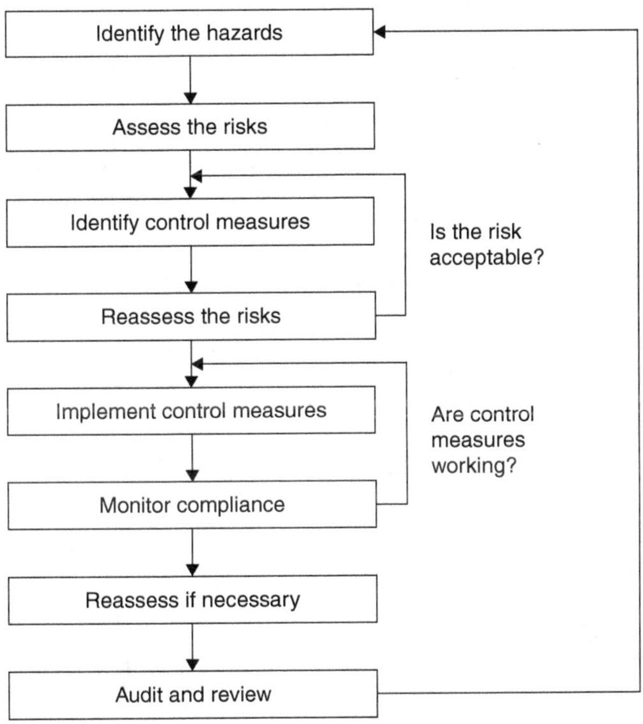

Figure 5.1 – Flowchart for risk assessment

almost as useless as no risk assessment. Figure 5.1 presents a generalized flowchart for risk assessment.

Principles of risk assessment

Risk assessment is determining the risk posed by an operation. In its most general form the risk of an operation can be described as the product of the consequences of a particular identified incident and the frequency of the particular incident happening. Commonly this is described as:

$$\text{Risk} = \text{Hazard (H)} \times \text{Frequency (F)}$$
Or
$$R = HF$$

To determine the overall risk of an operation each identified risk is summed for a variety of potential occurrences and thus consequences from a particular operation:

$$\text{Total Risk} = H_1F_1 + H_2F_2 + H_3F_3 + \ldots + H_nF_n$$

For example, as the result of a fire (from whatever source) in a magazine containing solely 1.4G fireworks, which are packaged and stacked properly, the overall risk comprises the factors listed in Table 5.1.

Table 5.1 – Typical risks from a fire in a magazine containing only 1.4G fireworks

Event	Hazard	Frequency
Rapid escalation leading to mass explosion	Building destruction, fragmentation, blast wave, "domino effects" to adjacent magazines	Very low
Projection of firework stars through open door	Burns, thermal effects, ignition of adjacent magazines etc.	Possible
Smoke plume, deposition of metal salts etc.	Toxic hazard to firefighters, environmental aspects etc.	Probable
Effects confined entirely within magazine	No hazard to outside; however, hazards during building clean-up etc.	Low

Safety and risk

Unfortunately, as practitioners of an occupation involving the use of hazardous materials we have to face two basic facts:

- The public are not good at statistics.
- The public do not understand the concept of risk.

Given these hurdles it is not surprising that the firework industry, and indeed any other industry that presents any form of risk to the public, battles to justify even their normal operation, and faces a barrage of criticism when something goes wrong.

Unfortunately we are not assisted by the law here. Most regulatory regimes require a person creating a risk to identify, assess and manage that risk — this is the basis of the often-abused Risk Assessment approach to safety management. This approach is the correct approach, although it simply is not realistic to consider hazard only, especially if hazard is considered in isolation from benefit.

Using a non-firework situation, for instance, the potential hazards from nuclear power generation are enormous — accidental meltdown after unauthorised experimentation, aerial contamination following a leak, long-term mutagenic effects from exposure to radiation, potential terrorist threat at the plant or from its products, massive waste-disposal issues, the potential rendering of large areas radioactive and contamination of the food chain — all are realistic potential hazards. What allows the world to even consider using nuclear power is firstly the benefits (electricity generation, lack of acidic or greenhouse gas emissions for such generation, independent security of supply), but more importantly the control of all the hazards by rigorous application of risk analysis and the resulting extremely low-frequency of any of the identified hazards arising.

The difference between risk and hazard

Hazard is the intrinsic harm that may be done by a particular sequence of events — for instance IF a shell bursts in the crowd, people will be injured or killed. It does not relate to the likelihood of that event happening — if a shell bursts in the crowd it makes no difference how the shell came to be there!

Risk relates the intrinsic hazard to the likelihood of that event occurring and thus equates the frequency of an event (or series of related events leading to the eventual hazardous outcome) with the hazard of that event to provide a perception of the risk. In essence, as has been seen,

$$\text{Risk} = \text{Frequency} \times \text{Hazard}.$$

In the case of the shell bursting in the crowd the intrinsic hazard is very high, but the likelihood is normally extremely low — and thus the perceived risk (and the calculated risk — see later) remains acceptable.

Unfortunately when a very rare incident does occur, the public and the press react as if that occurrence were actually commonplace, and worse, if the case comes to court, the court will tend to dismiss the frequency and hence the risk arguments and effectively state that in the particular case the frequency was one — it happened — and therefore the measure of the likelihood is now irrelevant and as a consequence only the hazard should be considered.

Until the discrepancy between the lawmakers and the courts is resolved, and until the public appreciate the difference, it is a battle that will continue to have to be fought — not only by the pyrotechnics industry.

Individual and societal risk

In considering risk we need to address the risk of several different classes of people:

- The risks to a single person in the audience or a display firer — usually taken to be the risks to a specified individual.
- The risk to a group of people, perhaps over an extended period of time.

Assessing the risk to a specified individual is both relatively straightforward and relates to the individual's experience of risk, whether they are benefiting from the activity causing the risk (e.g., they are a paid display operator), whether they might be aware of the risks being posed to them (again a display operator) and those who neither gain, nor can realistically be expected to understand the potential risks they are being exposed to (in most cases this is the situation of a member of the audience).

Assessing and justifying the risks to a group of people is much more difficult. Society has an understandable aversion to incidents that injure many people — which is not proportionate to the number of people injured. For example, we are generally much more averse to a road accident that kills ten people than we would be to ten accidents each killing a single person. This may appear callous, but it is a demonstrable fact borne out, not least, by the media reporting of such incidents.

Many authors have used the *f/n* curve approach[1] to attempt to quantify the societal risk to members of the public affected by an incident. In this approach *n* is the number of fatalities and *f* is the frequency of *n* or more fatalities, and the resulting plot may typically look like either a step function or a smoothed curve (Figure 5.2).

It is sometimes appropriate to attempt to superimpose on such a plot the public aversion to the risks identified, but there are also more sophisticated[2] ways of determining such aversion.

Societal risk[3] is difficult to quantify in a simple, unambiguous and objective way, and for the remainder of this section we will concentrate

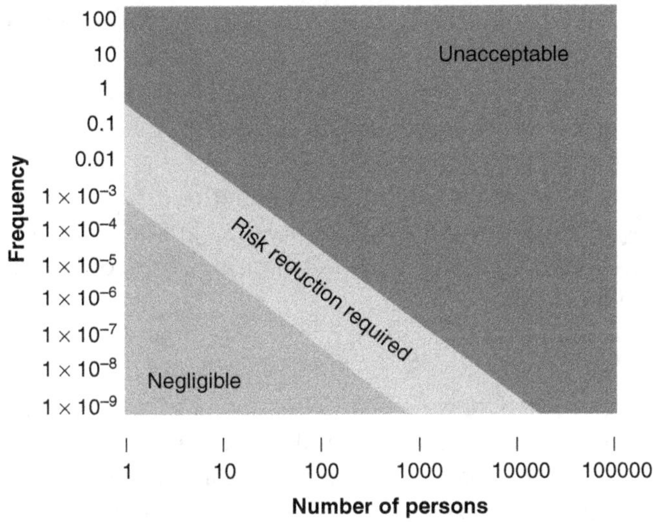

Figure 5.2 – Societal risk matrix

on quantifying individual risk. Where appropriate, however, illustrations of societal risk factors will be made.

Risks and benefits

At firework displays it is important to consider that alongside potential individual and societal risks there are individual and societal benefits associated with the display. For instance, for the firers and others involved in the event:

- Income for the firers
- Income generation for other performers and for those witnessing or benefiting from the event
- Getting people outdoors
- Income for charities at fund-raising events
- Socialising
- General individual "feel good" factor of witnessing a spectacular display
- Enjoyment of a spectacle by a very large number of people, often for no charge, at extended distances.

Just because there are associated societal risks doesn't mean that such events should be banned! There are plenty of examples where people, as groups or individuals, are prepared to accept significant risks for their benefit — even when the risks are not directly under their control — for instance:

- Flying away on holiday (risk of plane crash by mechanical failure or pilot error or by terrorist attack)
- Taking part in sports (risk of injuries or death)
- Watching a film in a cinema (risk of building fire or collapse)
- Driving (risk of crash caused by mechanical failure, or by driver error or by accident)
- Walking to a bar (risk of accident on the way, or as a result of drinking)

Any meaningful risk-assessment process needs to address the costs of identified risk-reduction measures, and thus to determine whether such costs are justified. Basing a paper Risk Assessment on unachievable or unaffordable control measures is not helpful or informative to anyone.

Cost-benefit analysis

For instance, in deciding if the risks arising from firing shells can be reduced it is necessary to consider:

- How many mortars are used on the display
- How many times they are reused (on the display and over their life-time)
- How many shows are fired in a year
- How much the control measure would reduce the risk
- How much each control measure would cost
- The cost that is placed on a life or on injuries

The last parameter may seem flippant, but without equating benefits and costs a risk-assessment approach is almost meaningless. It simply is not practical or sensible to consider that ALL identified control measures are justified. It may be possible to reduce the risk of fatality from, say, one in one million to one in two million — but is such a reduction statistically significant, and is the cost of such a control measure justified? If the cost is, for instance in the case of mortars, comparable to the cost of a mortar the answer is probably "yes" — if the cost is 100 times the cost of a mortar it is almost certainly "no".

In general we believe that any control measure that decreases risk by a factor of less than ten is probably not worth contemplating in isolation.

The simple mathematical treatment of risk

As we have seen, in simple terms, the specific risk of a particular event's causing harm is calculated as follows:

$$\text{Risk} = \text{Frequency} \times \text{Hazard}$$

The overall risk of a particular event is then the sum of the specific risks over all things (people or structures) affected by that risk, and ultimately summed over the extent and duration of the event (including the rigging, firing and de-rigging phases).

Low-risk events may occur if either the frequency or the hazard (or both) is low; the highest-risk events occur when the hazard and the frequency are high. In the simplest terms it is useful to illustrate the interrelationship between frequency, hazard and the resulting risk — this is shown in Table 5.2.

Table 5.2 – Very simple risk-assessment methodology

	Low frequency	**Medium frequency**	**High frequency**
Low hazard	Very low risk	Low risk	Medium risk
Medium hazard	Low risk	Medium risk	High risk
High hazard	Medium risk	High risk	Very high risk

In most cases we simply do not engage with high or very high-risk events at firework displays — it would be morally wrong, and commercially suicidal, to do so. However, as a consequence of only dealing with relatively low risks and especially with very low-likelihood events there is a danger of complacency on the part of the operator, an assumption that "low risk" equates to "safe" and often a reluctance to comprehend the actual risks posed.

Somewhat more complex approaches to risk assessment include ranking of both likelihood and hazard on a suitable scale and treating the product of the two numbers as the risk.[4] We generally use the approach where hazard and likelihood are both rated on a 0–10 biased scale — because the resulting product has a maximum value of 100, and because there are situations where either the likelihood or frequency is truly zero, and zero times anything is still zero! Biasing the scale puts greater emphasis on the highest-frequency or highest-hazard events and means, for instance, where a number of related events are considered together these high-frequency or high-hazard occurrences are not ignored.

The likelihood of an event happening is also rated on a 0–10 scale, also weighted to highlight highest-frequency events and to which we have tried to apply a rough frequency (F).

The RISK is then assessed by multiplying the LIKELIHOOD by the HAZARD. It is often beneficial to apply this calculation to both the USER and those AFFECTED by the risk, and to obtain an overall ranking.

The overall risks are then assessed by multiplying the two factors to obtain an overall risk ranking as shown in Table 5.5.

Table 5.3 – Hazard indices for humans and structures

Index	Effect on humans	Effect on structures	Examples
0	Nothing of consequence	Nothing of consequence	Ash on hand
1	Single trivial injury	Single superficial damage	Lit ash on hand causing very minor burn (e.g., from a sparkler)
2	Multiple trivial injuries	Multiple superficial damage	Minor scorch marks on structure — wash off with rain
3	Single minor injuries	Single minor damage	Ash in eye
4	Multiple minor injuries	Multiple minor damage	Structure requires repainting etc.
5	Single major injury	Single major damage	Loss of limb
6	Multiple major injuries	Multiple major damage	Structure requires replacement
8	Single fatality	Significant damage — partial collapse	Death — immediate or as a result of injury
10	Multiple fatalities	Significant damage	Total collapse of structure

Table 5.4 – Likelihood (frequency) of events

Index	Descriptor	Example	Indicative frequency
0	NEVER happens	Firework debris falling two miles upwind	$F = 0$
1	Very unlikely to happen		$F < 10^{-7}$

Index	Descriptor	Example	Indicative frequency
2	Happens only rarely	Shell explodes and bursts mortar	$10^{-5} > F > 10^{-7}$
3	Occasionally happens		$10^{-3} > F > 10^{-5}$
4	Happens	Firework fuse fails	$10^{-1} > F > 10^{-3}$
6	Frequently happens		$1 > F > 10^{-1}$
8	Almost always happens	Lit firework debris landing in firing area	$F \approx 1$
10	ALWAYS happens	Firework debris landing on ground	$F = 1$

Table 5.5 – Matrix of hazard and likelihood

	Likelihood (frequency)							
Hazard	0	1	2	3	4	6	8	10
	1	1	2	3	4	6	8	10
	2	2	4	6	8	12	16	20
	3	3	6	9	12	18	24	30
	4	4	8	12	16	24	32	40
	5	5	10	15	20	30	40	50
	6	6	12	18	24	36	48	60
	8	8	16	24	32	48	64	80
	10	10	20	30	40	60	80	100

The similarity to Figure 5.4 is clear!

And finally the results are assessed, for each of the risks identified, and to each of the areas affected.

	UNACCEPTABLE Risk – measures MUST be taken
	HIGH Risk – measures should be taken to control and reduce risks
	ACCEPTABLE Risk – seek advice where necessary
	VERY LOW Risk – adequately controlled

It is critical that the hazards or risks to all parties be considered and that a mechanism chosen to reduce the risk to one potential area does not adversely affect (without other controls) the risks to others.

As an example we are often told by firers that by firing the display electrically the risks to the operators are significantly reduced — however, what few people go on to recognise is that the risks to the audience may actually be increased. If the firing of one shell causes an adjacent mortar to be displaced, no one manually firing the second shell would light it — but if the firer is, perhaps, 100m away and is unaware that the mortar has been displaced, then he or she will (or the firing system will) fire the second shell — possibly directly towards the audience.

This is not to say, of course, that electrical firing is bad — but illustrates that the desirable consequences of a particular action may have unintended negative consequences, and that therefore all possible outcomes should be considered.

We tend to produce the Risk Assessment on a spreadsheet (Figure 5.3) that uses automatic cell formatting to highlight risks that are higher than acceptable. An example is given below.

The column headings are, for clarity:

- Item
- Hazard and Effect
- Site/date
- To whom
- Initial Risk
  - o Hazard Index
  - o Frequency Index
  - o Risk Product
- Minimise Risk by
- Managed Risk
  - o Hazard Index
  - o Frequency Index
  - o Risk Product

There are occasions, however, where the simplistic approach is not appropriate and a simple quantified risk assessment (QRA) leads to better understanding (and therefore potentially better mitigation) of the risks.

Safety distances

In developing the European Standards for "professional" fireworks and other pyrotechnics, the question of what is the "safety distance" is frequently posed by experts and enforcing authorities. Unfortunately, no

Risk assessment					Event	Display				
0-10 weighted scales - see key					Date	31/12/2005				
				Initial risk				Managed risk		
Item	Hazard and effect	Site/date	To whom	Hazard index	Frequency index	Risk product	Minimise risk by	Hazard index	Frequency index	Risk product
1	Premature ignition of items on firing plate during unloading	Site during rigging	Public	10	2	20	Keep public remote from unloading/transport route (via barge). Keep items packaged until on BALE	4	1	4
2	Premature ignition of items on firing plate during unloading	Site during rigging	Operators/staff	10	2	20	Only pyrotechnists/riggers to handle materials. BALE riggers to be supplied with fire-retardant PPE	10	1	10
3	Mass explosion of entire holding in transport vehicle leading to onsite risk	General	Any	10	3	30	Stock remain in transport boxes throughout	10	1	10
4	Mass explosion of entire holding in storage leading to off site fatality etc	General	Public	10	2	20	Maintain safety distances to barrier for public. No access to site for public. Consider crew/staff ID badges.	6	1	6
5	Unexpected violent event from ignition due to incorrect types	General	Operators/Public	10	2	20	Good housekeeping, labelling of boxes, clear instructions and monitoring	6	1	6
6	Ignition through poor handling of items/containers	General	Operator	10	2	20	Clear procedures from operator and monitoring	10	1	10
7	Protestor attack leading to compromising safety of site	General	Public	10	2	20	Adequate security	10	1	10
8	Accidental ignition during loading/unloading	General	Operators	10	1	10	Restrict numbers of personnel and stock at any one point in time	8	1	8
9	Premature ignition from smoking/matches etc	General	Operator	10	2	20	All sources of ignition banned from immediate pyrotechnic working site	1	1	1
10	Transport of fireworks from vehicle to working area	General	Operator	10	2	20	Boxes to be closed. Procedures to prevent excessive quantity/unsafe transport (eg overloaded trolley) being used	8	1	8
11	Adding electric fuses to piped match/firework body	General	Operator	10	3	30	Exposed fireworks to be kept to minimum. Igniters to be protected by sheath or similar. Majority of work completed offsite	6	1	6
12	Manipulation of blackmatch fuse	General	Operator on barge	10	2	20	Exposed quantity to be kept to a minimum	6	1	6

Figure 5.3 – Example of risk-assessment spreadsheet

activity involving pyrotechnics is ever "safe", not for operator and not for audience, and in many cases the nationally prescribed and often fixed "safety distances" may be regarded by others as overly draconian or too lenient. The major problem with fixed "safety" distances is that they do not provide a guarantee of safety — they tend to be arbitrary distances based on local custom and practice, and do not consider any risk control measures that have been adopted by the firer.

But in fact the question (and the answer) is wrong — what we really need to know is what fireworks can be fired at a given site, under the conditions prevailing and the methods and techniques adopted by the firer and hence provide an acceptable level of risk to those exposed to the risks. Any human activity poses some level of risk — it is just that we, both as individuals and as a society, tolerate some levels of risk more than others.

This section, adapted from an earlier article by the author,[5] illustrates the risk-assessment process for, predominantly, shells at a variety of sites under a variety of conditions, which leads to an assessment of the distances required from the point of firing to the audience under the specified conditions.

It does necessarily concentrate on the use of aerial shells at firework displays, as these inevitably pose the greatest hazards, but the principles employed would also be suitable for outdoor displays not involving shells, or for events using theatrical or other pyrotechnics.

The risks at firework displays

The risks at firework displays can be broadly separated into two areas:

- Risks from the normal functioning of the fireworks (e.g., expected debris)
- Risks from the abnormal functioning of the fireworks (by product failure or operator error)

Both are important in determining the choice of fireworks, the suitability of site and the inter-relationship between the two, as has been illustrated in previous chapters. If the site has inbuilt flexibility (e.g., a barge that can be moved to maximise the fallout area under any wind conditions), then the choice of fireworks for the display may be much more extensive than if the site is fixed (e.g., a rooftop). Furthermore, on the rooftop there may have to be various contingency plans to curtail or cancel the display in adverse conditions that may not need to be mirrored for the display on a barge.

If we take a typical firework display mix of products we can determine which are likely to pose the greatest risks, as shown in Table 5.6:

Table 5.6 – Potential risks from selected firework types

Firework type	Subtype	Potential risks
Shells	Colour shells	Normal debris (especially long-burning stars)
		Risks from sub-components
		Abnormal firing angles (from mortar disruption)
		Blinds (shells fail to burst)
		Blast/fragments from bursts
	Maroons	As above plus
		Increased blast/fragments from bursts
	Aqua shells	Unpredictable range
Roman candles	All types	Projectile effects
Rockets	Flight rockets	Random flights
	Display rockets	Blast/fragments from burst
		Sticks falling to earth
Mines	All types	Minor projection effect
		Subcomponents
Fountains	All types	Sparks
		Failure of casing — unintended explosion
Set-pieces	Lancework	Relatively minor
	Wheels and set-pieces	Throwing driver from wheel
		Failure of support

The potential risks from "normal" shell debris are relatively low (although the frequency of "normal" debris falling on the audience is surprisingly high, the hazard is low and hence the risk is low), and most firework companies have developed tables equating shell-burst height and wind strength with fallout distance.

The Shellcalc© programme, developed by Harradine and revised by the author and discussed in detail later in this book, calculates the trajectories of shells and comets depending on a variety of user-inputs, including:

- Calibre
- Firing angle
- Wind direction and strength

- "Barrelling" or "tumbling" effects (which add a realistic variation in shell dispersion)
- Muzzle velocity (if known)
- Mass of the shell or comet if known (and allowing that comets having a decreasing mass during flight)
- Shell delay (for determining if the shell reaches the ground or bursts in the air)

In general, Shellcalc© is not particularly useful in determining "normal" debris from the "normal" functioning of shells. For this, analysis of wind speed and direction from the burst point (which of course may itself be estimated by using Shellcalc©) and the time for fallout to reach the ground (or other areas) are more useful. This is discussed further in Chapter 7.

In the case of shells, in general it is the largest-calibre shells, fired from displaced mortars, that pose the greatest risks. However, lower-hazard, higher-frequency events may actually pose the same or at least significant risks to both operators and the audience. For instance, the main "normal" debris from rockets (their sticks) can travel significant distances downwind of the firing site and has the potential to cause significant harm.

For a display with several different calibres of shells, which is the norm, the high-hazard/low-frequency failure of the largest-calibre shells may be outweighed by less hazardous but more frequent failures of lower-calibre shells. This arises for four main reasons:

- In general, large-calibre shells are manufactured to a higher quality than small-calibre shells (for instance, they may contain dual or multiple internal delay fuses).
- There is usually a higher number of smaller-calibre shells than large-calibre shells in a display — indeed the numbers of shells fired are usually inversely related to their calibre (because of cost and aesthetic features).
- Smaller shells are more usually fired at greater deliberate angles than large-calibre shells.
- Smaller shells are more usually fired from "racks" than larger-calibre shells.

However, as will be seen, the risks from the largest-calibre shells are usually the ones considered in risk assessments. This is because

- Larger shells generally rise to a greater height — therefore "normal" debris is likely to travel farther downwind than debris from smaller-calibre shells.
- If the mortar is disrupted the range of the shells is greater.
- The burst charges of larger-calibre shells are greater and hence likely to cause greater injury if a larger-calibre shell bursts in/near the audience.

Mathematical evaluation of risk

The remainder of this section will address a more complex mathematical evaluation of the risks posed from shells to the audience at firework displays — a so-called "Quantified Risk Assessment" (QRA). The principles applied are also valid for other firework types, and for use of other pyrotechnic devices (for instance indoors).

To evaluate the risks to individuals in the audience from shells, we first need to consider what are considered the benchmarks for the acceptability of risk.

In general the following are the UK accepted guidelines[6] for risk of fatality:

- 1×10^{-6} — broadly acceptable
- 1×10^{-5} to 1×10^{-4} — the so called "ALARP" region (1×10^{-5} to 1×10^{-3} for workers)
- $>1 \times 10^{-4}$ — unacceptable ($>1 \times 10^{-3}$ for workers)

This means that a member of the public should consider broadly acceptable the chance of being killed at a firework display as one in a million — about ten times the chance the same individual has of winning the UK lottery in any one week. This seems at least reasonable — no one expects to win the lottery, and no one should expect to be killed at a firework display!

In general we will use the 1×10^{-6} "broadly acceptable" criterion for determining the risk to an individual.

The so-called "ALARP" region (discussed below) is where the level of risk is still permissible, but where measures should (if at all possible and practical — especially when related to cost) be implemented to reduce the risk still further.

It is accepted that people who directly benefit from an activity involving risk (i.e., firers) may be subject to a greater range of ALARP than would the audience, and this is reflected in the figures given above. However, neither should ever be exposed to risks in the "unacceptable" region.

ALARP – "As Low As Reasonably Practical"

The UK Health and Safety Website[7] includes the following observations on the ALARP principle:

> *[D]etermining that risks have been reduced ALARP involves an assessment of the risk to be avoided, of the sacrifice (in money, time and trouble) involved in taking measures to avoid that risk, and a comparison of the two.*
>
> *This process can involve varying degrees of rigour which will depend on the nature of the hazard, the extent of the risk and the control measures to be adopted. The more systematic the approach, the more rigorous and more transparent it is to the regulator and other interested parties. However, duty-holders (and the regulator) should not be overburdened if such rigour is not warranted. The greater the initial level of risk under consideration, the greater the degree of rigour HSE requires of the arguments purporting to show that those risks have been reduced ALARP.*

It is clear that the costs are critical in determining the proportionality of any risk-control measures. Demonstration, by means of analyses such as presented in this section, should be enough to satisfy that the risks have been reduced to ALARP and hence that they should be accepted by operators, the audience and, if necessary, the courts —provided, of

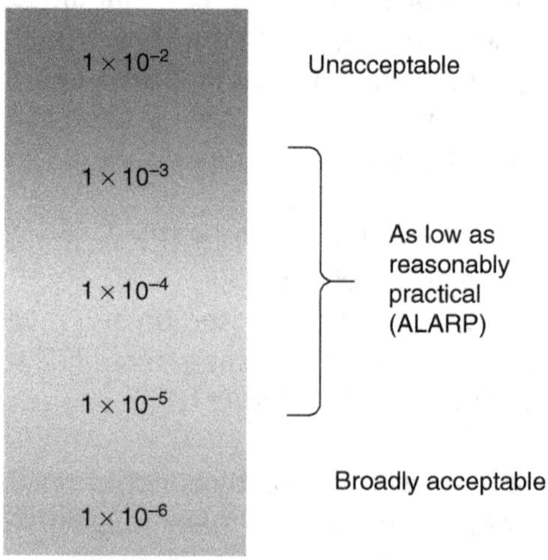

Figure 5.4 – The ALARP diagram

course, that they truly represent the risks involved and that operator error or disproportionate ratios of product failure have not been a contributing factor.

Risk to the operator vs risk to the audience

In all assessment of risks it is essential to consider all consequences of the identified hazard. For instance, in firing shells it is important to consider:

- The effects on the audience AND
- The effect on the operators

Some things that reduce risk for operators MAY increase risk to audience (and vice versa). For instance, as outlined above it is often (correctly) stated that electric firing of shells is safer for the operator — the operators are physically removed from the mortars and have no possibility of positioning any part of their bodies over a loaded mortar tube. However, it possible that electric firing of shells actually poses a greater risk to the audience than manual firing. It is conceivable that the firing of one shell (as discussed in detail below) displaces an adjacent mortar so that it is directed at the audience — if the firer is "merely" pushing a button 100m away along a piece of wire they will fire the second shell unaware that it is lying in a displaced mortar. A person manually firing would simply not fire the second shell! This is NOT to say electric firing is bad — just that the consequences of one risk-reduction method might actually increase the risks to another party and therefore appropriate measures to control the additional risk should be taken.

For simplicity, for the remainder of this investigation we will concentrate on the effects to the audience only.

Catastrophic failures

Low-frequency (or very low-frequency) events with catastrophic outcomes are one of the most essential areas to consider in risk assessment. For most displays such catastrophic failures are likely to involve projected effects — especially shells. This can occur in several ways:

- A shell being projected "normally" and failing to burst in the air, and then reaching the ground
- A shell being projected "normally" but in an unintentional direction — so that it is projected towards the audience or other hazards — for example, when an adjacent firework malfunctions

- A shell "detonating" in a mortar, causing mortar fragments to be projected
- A shell bursting near to ground level and throwing stars into the audience

Hazards from shells

There are several major hazards from the firing of shells as shown in Table 5.7. Of course, the risks arising from these hazards depend on the frequency of each event occurring. In the remainder of this paper we will concentrate on fatality hazards, and risks to the audience.

Table 5.7 – Potential hazards from shells

Type	Hazards	Comments
"Normal" functioning of a shell	Lit debris drifting downwind, especially from long-burning stars	Generally unlikely to cause fatalities by direct action — although may have impact on structures, leading to fire
"Abnormal" functioning of a shell when fired in designed orientation	"Blind" shells	Obviously worse if the mortar is angled
Premature functioning of a shell	Flowerpot, muzzle break or in-mortar explosion — especially if it leads to disruption of adjacent mortars	See below.
Disruption of the mortar from external event and subsequent "normal" firing	Low-bursting shell at unplanned firing angles — risk if stars reach audience	If mortar is fired near vertical, the risk of stars impacting the audience is very low.
Disruption of the mortar from external event and subsequent "abnormal" functioning (e.g., "blind")	Shells fired at unplanned angles — risk if shell lands in or adjacent to audience	Impact or close-proximity effects from fragments and blast

It is important to identify the critical events to consider, and where possible, to separate key events leading to various scenarios.
For example:

- What is the overall rate of shell failures (of any type)?
- What proportion of general failures could affect the correct functioning of an adjacent mortar?

- If a mortar is fired at an undesirable (and unplanned) angle, what proportion of those could affect the audience (i.e., are "shot" towards the audience)?
- What proportion of shells fired from disrupted mortars could actually impact the audience, and will function "normally" without affecting the audience — for instance, by bursting at a sufficient height so that the stars do not reach the audience?
- What proportion will function "abnormally" (e.g., fail to burst in the air) and thus present a different hazard to the audience?
- How many people may be affected by each failure mode?
- What is the frequency of injuries/fatalities from the postulated shell-failure mode?

We have made some very general assumptions on the basis of data regarding shell failures collated from informal and formal surveys of the UK firework industry.

For injury and fatality data we have examined the UK Explosives Incidents (EIDAS)[8] database for reported accidents involving shells. The results are shown in Table 5.8.

Note that in several cases identified in EIDAS it appears that fragments of the mortar were the cause of the injuries/fatalities — rather than the shell itself. This is obviously an important factor, particularly when using

Table 5.8 – UK shell accidents from EIDAS database

Date	Location	Injuries	Fatalities	Cause comments
5/11/2005	Kettering	1 major, 10 minor		Rack collapse, shell burst near crowd
1/11/2004	Middleton	1		Shell fired into leg
3/11/1996	Wilmington		1	Incorrectly fired (Cat 3)
2/11/1996	Hazlemere		1	Incorrectly fired (Cat 3)
1927	Unknown		1	Head over mortar
25/6/1910	Leeds	7	3	Shell detonated in steel tube
19/9/1898	Folkestone	1		Mortar burst
2/6/1896	Doncaster	3	1	"Mortar" lands in audience
5/8/1895	Brighton	25	1	Mortar burst
24/12/186	Batley	5	1	Mortar tipped over prior to firing
25/12/1882	Hull	3	2	Mortar burst
19/7/1882	London		1	Mortar burst

metal mortars, but we have not considered it further here for three main reasons:

- The use of metallic mortars (particularly steel) is decreasing.
- The injuries are most likely to occur to operators, not the audience.
- There has been a general trend to move fireworks farther from the audience for aesthetic and practical reasons, as well as for generally perceived safety reasons.

It is also relevant to note that the number of accidents is actually very low — the data span 1882 to 2005 and include the period (up to 1996) when shells were available for the general public to purchase and use.

Where a shell lands in the crowd, the apparent outcome is most likely to be a single fatality (if a fatality occurs), so we have set the likelihood of such a fatality to be 1. For shells that do not burst in the crowd we have set the fatality likelihood as 0.1 (see Table 5.9).

Table 5.9 – Failure modes of shells involving fatalities

Failure mode	No likelihood of fatalities	Comments
Shell bursts above the audience so that stars reach the audience.	Multiple injuries Single fatality (hazard = 0.1)	
Shell bursts in or near audience so that people are affected by impact or by bursting of the shell or by fragments of the shell.	Multiple injuries Single fatality (hazard = 1)	It is possible for multiple fatalities to occur in this situation — a brief examination of the societal risk aspects will be made below.

All the above information is collated in Table 5.10:

Table 5.10 – Collated frequencies/hazards from shell failures

Failure	Estimated frequency	Comments
Failure of a shell in any manner	0.01	This is pessimistic and improvements in manufacture are reducing this. However, this figure is given for ANY failure of a shell.

Failure	Estimated frequency	Comments
Above shell leading to mortar failure	0.01	
Disruption of adjacent mortars	1	We have assumed any failure of a mortar will cause disruption of adjacent mortars in a rack or trench. This is overestimating significantly.
Adjacent tube contains a shell	0.5	The adjacent tube may be empty or non-existent (failure of last shell in a rack). The evidence from accidents (e.g., Kettering) suggests some shells will remain unfired in adjacent tubes even if disrupted.
Angle factor — adjacent shell fires towards audience	0.1	This is variable — see text.
Fatalities from shell bursting above audience	0.1	Measure of hazard to person standing within burst radius of shell
Fatalities from shell bursting in or adjacent to audience	1	Measure of hazard to a person standing within the immediate burst area where they will be affected by blast
Fatalities from fragments of a shell bursting adjacent to audience	1	Measure of hazard to person standing in immediate burst area of a shell where they will be affected by fragments
Frequency of shell bursting adjacent to mortar	0.01	For instance, muzzle break or flowerpot. In most cases affects operator only.
Fatalities from shell bursting adjacent to operator	0.1	

The nature of the display site and the scale of the display also affect the likelihood of a particular shell reaching the audience.

Most significant is the angle that the audience subtends (i.e., occupies), at the perimeter of a circle drawn around the display site. Examples are given in Table 5.11. It is often the case that smaller professionally fired displays subtend relatively small angles to the audience — a private function, for example, is likely to have a relatively small audience in a relatively small area.

Table 5.11 – Subtended angles at shows

Type	Description	Typical angle subtended
Where the audience only subtends a small fraction of a circle around the firing site	Typically an event where the crowd is small and well controlled — e.g., a wedding where the audience is assembled on the steps of a hotel	36° 0.1 of a circle
Where the audience is on one side of a display site	Typical of many shows	180° 0.5 of a circle
Where the audience is all around the display site	Large displays, or displays from rooftops	360° A complete circle

Larger displays often are fired from positions where the audience may subtend significant angles — up to situations where the audience, in effect, surrounds the display site. However, in general, as the display gets larger the proportion of shells that can reach the audience decreases; the audience is beyond the design range of the smaller calibre shells, and hence the risks from the smaller shells bursting in the crowd are reduced to near zero. This can be used to calculate the likely risks in one of two ways:

- By only performing calculations on the largest shell calibres and counting only those largest-calibre shells
- By applying an approximate "show factor" for the show — but counting all shells fired — for instance as shown in Table 5.12. Such "show factors" should be developed by individual display companies to adequately reflect their particular style of show designs.

We have used both methods and find they yield similar results. However, where the normal pattern of distribution of shell sizes is inappropriate,

Table 5.12 – Typical show factors applied to the total number of shells fired in a display

Type of show	Shells used	Typical value of show factor
Small e.g., private wedding	Several 75mm Several 100mm 1 × 150mm	0.2

Type of show	Shells used	Typical value of show factor
Medium e.g., public concert	Many 75mm Many 100mm Many 125mm Many 150mm Few 200mm	0.1
Large e.g., national event	Very many 75mm Very many 100mm Very many 125mm Many 150mm Many 200mm	0.05

calculations may have to be made across a variety of shell calibres/ numbers and the results combined to give a value of the total risk.

Proportionality factors

HSE in the UK recognise that any work required to mitigate a particular risk must be proportionate to the reduction in risks achieved. The higher the original (unmitigated) risk, and the greater the risk reduction, the more affordable are the mitigation measures.

For instance, taking the "cost" of a life as £1 million, a reduction in risk from the "broadly acceptable" (1×10^{-6}) to lower would be justified if the benefits exceeded £1 million (i.e., there is a proportionality factor of 1 in this case). For a risk of roughly 1×10^{-5} the proportionality factor would be arbitrarily set at about 4, i.e., measures would be justified if the benefits exceeded £250 000 or in essence one-quarter of a life, or significant numbers of major injuries.

The costs associated with such benefits can be spread over many mortars and many uses of the mortars, and effectively amortised over the lifetime of the mortar. For instance, for a medium-sized company

- Doing 100 displays each year
- Holding 2000 mortars in stock
- An average number of shells fired per show is 125
- With a total number of shells fired over a 4-year period of 50 000
- Where each mortar is reused 6.25 times a year and 25 times in a 4-year lifetime

if we assume that shows have an audience of 1000, of which 100 are at risk and the audience subtends 1/10 of a circle, then examining four possible risk-reduction measures should illustrate the measures that are sufficient or necessary:

- Tinfoil over mortars to prevent ignition from stray sparks (for instance, if the mortar is disrupted by an adjacent mortar explosion)
- Waterproofing mortars to boost structural integrity (e.g., for fibre-board or GRP mortars to prevent freeze-cracking)
- Replacing each mortar each year
- Redesigning and implementing new mortar racks

For a large display company firing multiple displays on the same site (or a very large display for a single event) the calculations are somewhat different. Assuming that the display is as shown below, this leads to a subtly different set of conclusions:

- The display is repeated ten times each year.
- 2000 shells are used per display (or 20,000 per year).
- Multiple shows have a 100,000 audience, of which 1000 are at risk and the audience subtends at least 1/2 of a circle, or a single very large display has a 1 million audience subtending a full circle.

Where there are many displays each year (for instance at theme parks) more elaborate control measures may be justified, but in general some measures are justified for all shows and relatively simple "housekeeping" is essential for all. Mortars and racks should be checked regularly to ensure the integrity of both.

The results are shown in Tables 5.13 and 5.14.

Table 5.13 – Risk reduction for medium shows

Risk-reduction method	Cost per mortar	Total cost over 4-year period	Estimated risk reduction	Cost/benefit analysis indicates control measure worth doing
Tinfoil to protect shells from sparks	£0.10	£5000	5×10^{-6} to 1×10^{-6}	Yes
Waterproof mortars (renew each year)	£1.00	£8000	5×10^{-6} to 5×10^{-7}	Yes

Risk-reduction method	Cost per mortar	Total cost over 4-year period	Estimated risk reduction	Cost/benefit analysis indicates control measure worth doing
All new mortars each year	£10.00	£80000	5×10^{-6} to 2.5×10^{-7}	No
Redesigned mortar racks (each with 5 mortars)	£20.00 (i.e., £100.00 per rack)	£40000	5×10^{-6} to 1×10^{-7}	No

Table 5.14 – Risk reduction for large shows

Risk-reduction method	Total cost for 10 shows (or single very large show)	Estimated risk reduction	Cost/benefit analysis indicates control measure worth doing
Tinfoil to protect shells from sparks	£2000	5×10^{-5} to 1×10^{-5}	Yes
Waterproof mortars	£20000	5×10^{-5} to 5×10^{-6}	Yes
Additional sand barriers, new racks etc.	£200000	5×10^{-5} to 2×10^{-6}	Marginal — may be justified for large-budget productions
"Catchers"	£500000	5×10^{-5} to 1×10^{-6}	Generally no

Disney-style catchers are barriers erected to stop low-trajectory shells reaching the audience — for more information see Chapter 8.

The risks from firing shells range from "broadly acceptable" to lower risks — they DO NOT pose unacceptable risks. Larger displays merit greater in-depth analysis than smaller shows and can justify additional expenditure on risk-reduction measures.

Individual risk model

This section looks at the use of Shellcalc© data to investigate the individual risk to persons in the audience of a firework display. This approach could be applied generally, but for the purposes of this section we have made the following assumptions for inputting data into Shellcalc©.

- Ignore wind.
- Factor in "typical" tumbling and/or barrelling.
- Assume standard mortars, shell weights etc.
- Run the Shellcalc© model using 150mm shells.

We have also made the assumption that if a mortar falls over, there is an equal likelihood of the shell firing at any time, and therefore at any angle, as it falls to the ground.

We have then calculated the range and likely effect of firing a shell towards the crowd at 5° increments and investigated whether a shell

- Bursts over the audience at such a height that the stars impact on the audience
- Bursts at or near ground level so that persons may be affected by blast or fragments

The raw data derived from Shellcalc© are given in Table 5.15.

Table 5.15 – Data for a 150mm shell derived from Shellcalc©

Angle of firing	BLIND burst range	Normal burst range	Normal burst height	Stars reach the ground between the distances.	
0					
5					
10					
15	188	118	207		
20	226	147	195		
25	259	173	180		
30	286	199	162		
35	309	222	143		
40	325	243	121		
45	334	262	98		
50	339	278	74	270	290
55	336	292	49	240	350
60	327	303	22	230	370
65	310	310	0	230	390
70	282	282	0	210	360
75	226	236	0	180	310

Angle of firing	BLIND burst range	Normal burst range	Normal burst height	Stars reach the ground between the distances.	
80	109	169	0	90	250
85	33	33	0	– 40	110
90	11	11	0	– 80	80

For more details of the Shellcalc© program see Chapter 7.
From these data we have applied the risk calculations outlined above to calculate the individual risk for persons standing at distances of 0 to 500m from the firing point, relating to

- Very early bursts (e.g., muzzle breaks)
- Displaced mortars firing shells that burst "normally"
- Displaced mortars firing shells that function abnormally

In each case we also examine hazards from

- The effects of burst at very close distances (2m — taken as fatal at this distance)
- The effects of fragments of shells at extended distances (10m — taken as fatal at this distance) — this, in general, combines with the burst hazard, so anyone within 10m has a fatality frequency of 1
- The effects of stars (taken to the burst radius of the shell — roughly 80m — with a fatality of 0.1)

The risk calculations relate to the number of shells fired according to the formula

$$Risk_n = 1 - n(1 - Risk_1)$$

and not simply

$$Risk_n = n \times Risk_1$$

where $Risk_n$ is the risk from n shells and $Risk_1$ is the calculated risk from a single shell. Tossing a coin 10 times does not mean you will get 5 "heads" (10 × 0.5 = 5) — it doesn't even mean you will get any "heads". The likelihood you will get at least one head in this case is 1 — (1/1024) or approximately 0.99. For very small probabilities the two formulae tend to coalesce.

Figure 5.5 shows the distribution of distances where either a "blind" shell may fall or the stars from a "normally" functioning shell might reach the ground. The distribution was calculated by analysing the Shellcalc©-derived data in an Excel spreadsheet and analysed according to the following criteria:

- That if a mortar is displaced there is equal probability of it firing at any angle of displacement
- That a "blind" shell is considered to affect an area of its impact point on the ground and 10m either side of it
- That the analysis is carried out using 10m increments of distance — up to 500m from the firing point

The likelihood of a shell failing in either mode is then applied to the range distribution, and the overall risk calculated.

The risk is then rebased because the Excel analysis over-counts shells — it effectively counts all possible times a shell affects the audience, whereas the audience will only be affected once per single shell.

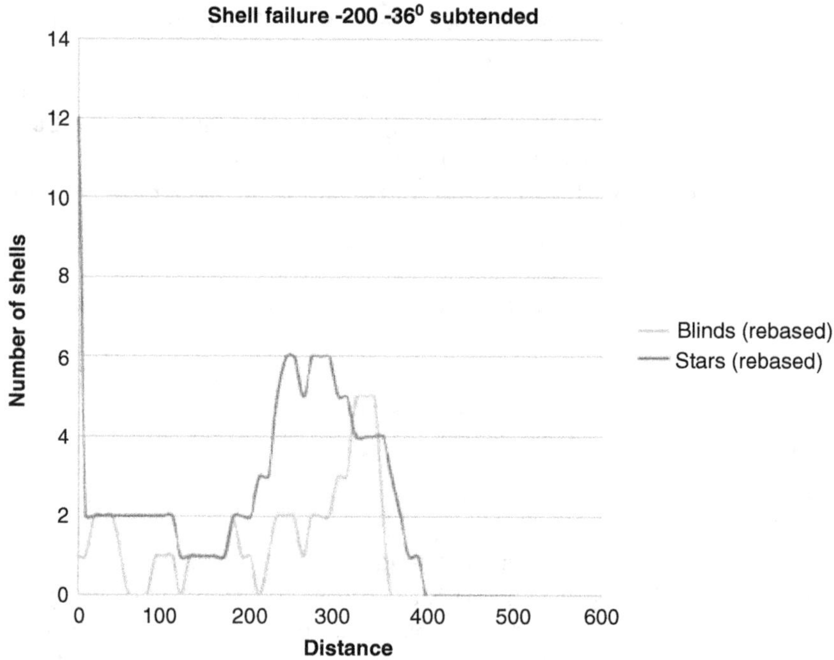

Figure 5.5 – Shell failures (blinds and stars) against distance for 200 shells fired with an audience subtending 36°

Figure 5.5 does not imply that, for instance, five shells reach the ground at approximately 250 m; rather, it implies that if stars from a "normally" functioning shell reach the ground they are five times more likely to reach the ground at 250m than they are at 150m. This is entirely in accordance with real-life observations and the Shellcalc© plots. If a shell is fired at a small deviation from the vertical and functions normally, then the stars do not reach the ground.

The "jagged" nature of the plot reflects the mathematical analysis used, and does not mean that there are particular distances where the likelihood of a "blind" shell falling or the stars reaching the ground is particularly high or low. We have chosen to ignore this "jagged" anomaly.

The plot for a show containing 200 150mm shells and subtending an angle of 36° (i.e., 1/10 of a circle) is shown in Figure 5.6.

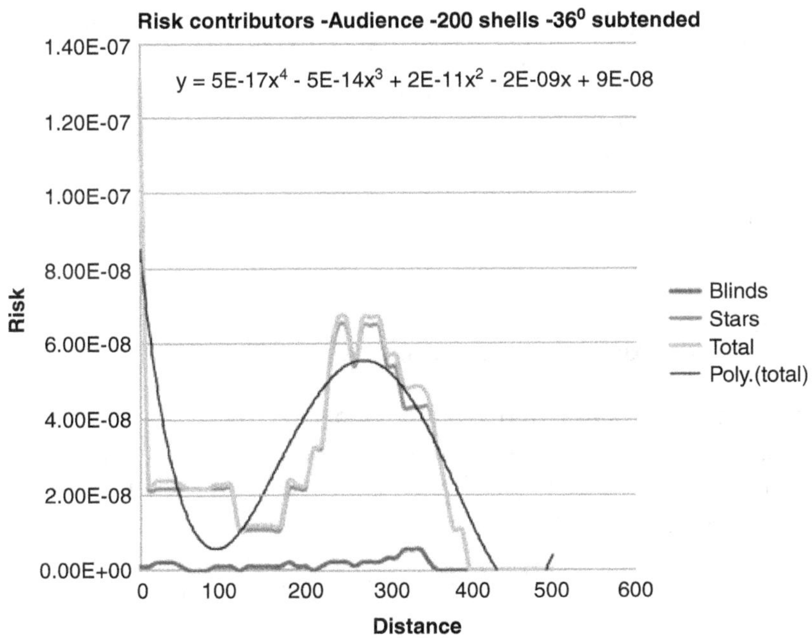

Figure 5.6 – Risk contributors — 200 shells, 36° subtended

Excel also allows an estimated regression to be made, which is also shown in Figure 5.6, and is a more realistic evaluation of the risk at any particular point.

A similar plot for a show containing 500 shells where the audience subtends 360° is shown in Figure 5.7.

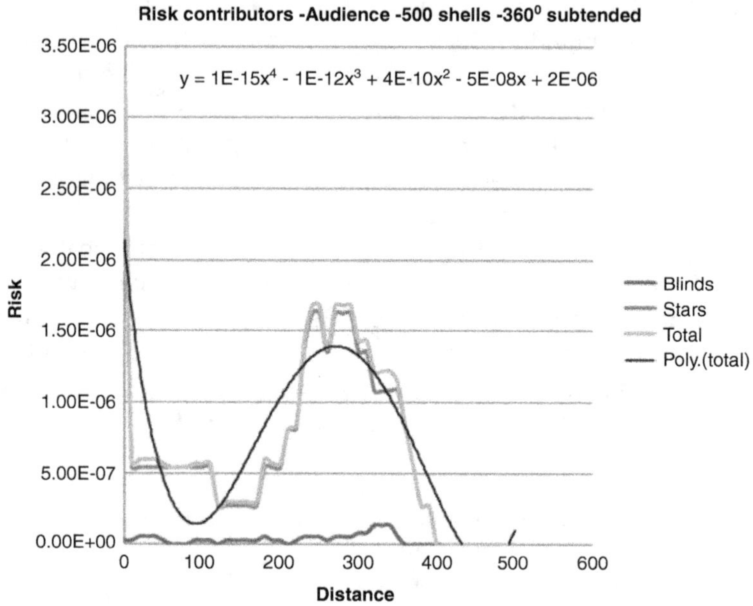

Figure 5.7 – Shell failures — 500 shells, 360° subtended

From these plots it is possible to calculate

- The distance at which an individual is subject to more than a "broadly acceptable" — i.e., 1×10^{-6} risk
- The total risk for persons at all distances

In the case of 500 shells fired at 360°, the distance at which an individual is subject to the same risk is approximately 360m. There is an area nearer the firing point at which the risk is the same — but this is discounted for obvious reasons.

Overall the risks remain extremely low — even in the second case the risks do not approach twice the "broadly acceptable" risk at any point.

Fallout

Addressing issues from fallout is critical for any display, and hence this topic will be covered in detail in a following chapter. However, the essential things to recognise are:

- Almost all fireworks produce fallout of one or more types.
- Different fireworks produce different types of normal debris when the firework functions.

- Different fireworks produce different types of fallout if they fail to function as intended.
- Very often the fallout may be hot.
- The strength and direction of the wind have a very major effect on where this fallout lands.

Resources for event planning

One of the most important things to determine on-site, once a full understanding of the risks posed by the chosen fireworks is established, is to have accurate information about the distances on-site. There are various ways of measuring the distances including:

- Pacing out the site
- The use of tapes
- The use of range finders
- Measuring from publicly available maps (e.g., Google Maps© or Google Earth© or in the UK MAGIC[9] maps)
- Measuring from site- and event-specific plans (e.g., CAD drawings)

In general, the progression above mimics the complexity of the display — highly detailed CAD plans are usually only available for the largest displays. In addition, the progression above also reflects the accuracy and precision of the measurements — although it should be noted that accuracy is not the same as precision; after all, fireworks are rarely precise in their functioning.

Google Earth© is particularly useful to get an overview of the site and to identify local hazards, but it should not be used as a substitute for a site visit and proper site planning — it is not that, in general, its accuracy is in doubt (although there are some well-documented areas where the mapping is wrong) but that it reflects the situation at a particular point in time (often quite a long time ago) and significant developments may have taken place "on the ground" that will not be reflected in the current Google images. Of course this is a problem with all maps — but it is also a potential problem when physically visiting a site.

- The site may change between the visit and the display date.
- Potential hazards are difficult to identify from ground level — local knowledge can be invaluable.

Figure 5.8 – Google Earth view of the Brockham Bonfire site

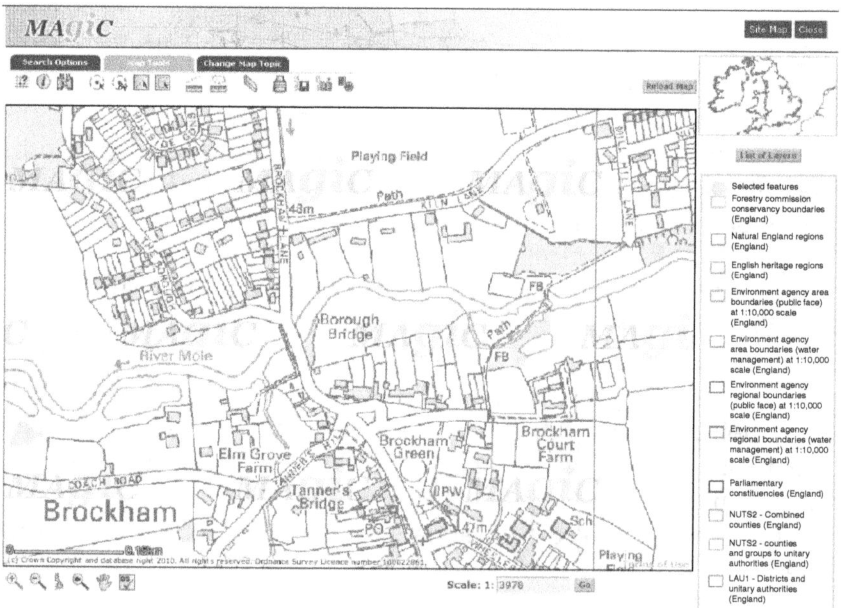

Figure 5.9 – MAGIC map of approximately the same area

Other "free" resources are also available, including the following:

Table 5.16 – Internet mapping software

Name	URL	Comments
Google Earth	Stand-alone programme available from www.google.com	There are some anomalies in the data (check for instance the Plymouth Breakwater), and care must be taken to ensure distances measured are not affected.
Google Maps	www.google.com/maps	
MAGIC	http://magic.defra.gov.uk/	UK maps only
Bing maps	http://www.bing.com/maps/	
Ordnance survey information	http://www.ordnancesurvey.co.uk/	UK maps
National map	http://nationalmap.gov/	US maps

Weather information

The final piece of information needed by the display planner (and the event organiser) is an appreciation of the likely weather conditions on the day of the event. It is no good visiting a site in May and expecting the weather to be the same in November! The important things to consider are:

- What is the likely weather on the display date — i.e., is it winter or summer?
- What is the prevailing wind direction at the site? This may be subject to quite local effects such as funnelling between buildings or up a valley — and these may be the "most likely" conditions.
- What are the "worst-case" wind conditions (perhaps towards the audience or another hazard)? Occasionally (but undesirably) this can be the same as the "most likely".

A working knowledge of meteorological charts, and in particular an understanding of high- and low-pressure weather systems, can be invaluable.

Weather (historical and predicted) and wind information is available from a variety of sources (Table 5.17).

Table 5.17 – Weather information

Information source	Application and comments
Wind data from local airports	Airports often have extensive historical information and accurate up-to-date observations. Wind "rose" data can be useful — but no historical data can predict what will happen at the actual time of the display.
Local amateur meteorologists	Many amateurs now have wind information published on the Internet — local gliding and sailing clubs are a particularly useful source.
National data	National data are available from government but can often be expensive and, as stated above, historical data are interesting but are not a predictor. They can demonstrate "most likely" and "worst-case" frequencies.
Weather forecasts	Most countries now have reasonable weather forecasting available via the Internet, but these are often relatively general and do not account for local features.
GRIB data	GRIB (GRIdded Binary) is a mathematically concise data format commonly used in meteorology to store historical and forecast weather data. The website www.grib.us allows users to download and display GRIB data for any area of the world.
Specialised websites	There are a number of specialist websites — details below.

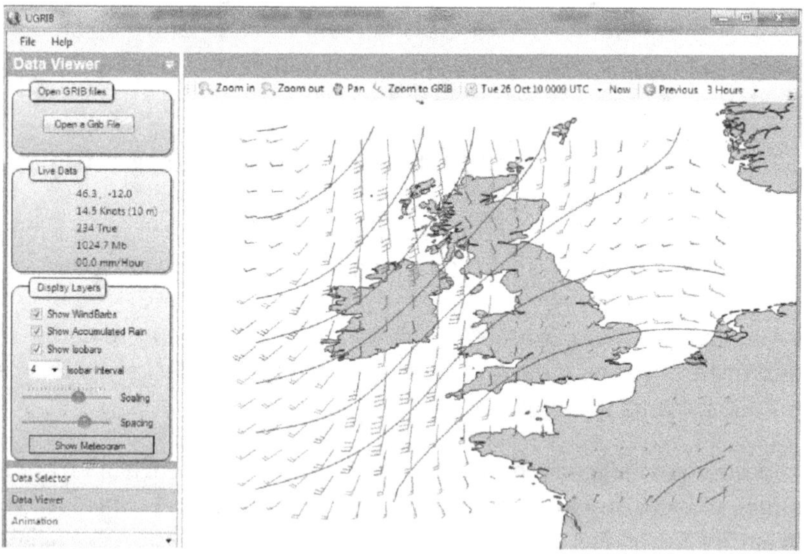

Figure 5.10 – Sample GRIB.us data for the UK

Table 5.18 lists some of the common websites for obtaining weather information.

Table 5.18 – Internet weather sites

Name	URL	Comments
GRIB.US	www.grib.us	Worldwide data
XC Weather	www.xcweather.co.uk	European data — provide prediction up to 7 days in advance
BBC Weather	www.bbc.co.uk/weather	UK data
Met Office	www.metoffice.gov.uk	UK data
National climate data centre	http://www.ncdc.noaa.gov/oa/ncdc.html	US data
NRCS Windrose data	http://www.wcc.nrcs.usda.gov/climate/windrose.html	US data
NOAA	http://www.weather.gov/	US data

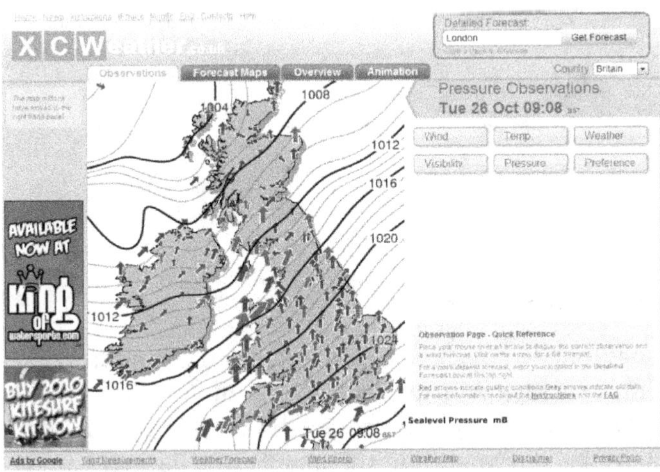

Figure 5.11 – Pressure chart for the UK from the XCWeather website

Finally, specific data may be obtained from national monitoring stations or local airports. This is often most usefully portrayed as a "wind rose" that indicates the most likely wind strength and direction.

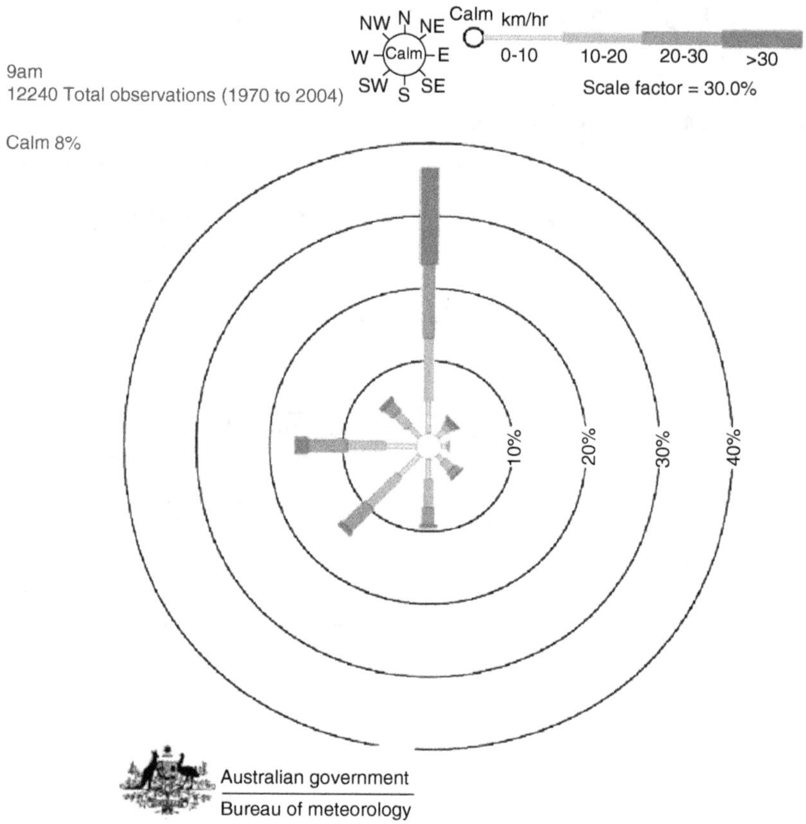

Figure 5.12 – Wind rose data

The health and safety culture

It would be wrong to conclude this chapter without addressing the concerns of many firework users that the "Elf 'n' Safety" culture has impacted the fireworks industry so much that it is almost impossible to perform a display anywhere but the middle of a desert.

Fireworks are explosives, and they have the potential for great harm if manufactured, stored, transported or used inappropriately. All these aspects (and more) are highly regulated, and in the main, the regulations are adhered to by responsible members of the industry. However, there is a significant anti-firework lobby that has corrupted the public's, enforcers' and politicians' minds to an extent that it is inconceivable that controls on fireworks will be relaxed. This isn't a UK or US problem — it is a worldwide problem.

Of course there have been accidents, some of which are devastating both to the individuals concerned and to the surrounding population and to the environment. But these accidents are rare, especially in relation to the number of fireworks produced, the number of transport journeys made and the widespread use of fireworks around the world. This is not to imply complacency, just to put matters into perspective!

The basic principles outlined in this book, together with an expectation that those who read the book will at least try to do things correctly, will assist those involved in all stages of firework displays to minimise the likelihood of any mishap occurring and to reduce the consequences of such an event.

Enforcers too have a role though, and under many jurisdictions they also have a responsibility (even if they do not like it) to ensure that where they are involved in the planning process they have sufficient knowledge to ensure a positive outcome from their enforcing activities. Too often we have seen well-meaning but technically ignorant people making outrageous proposals about how displays should be organised, set up and fired. Often this ignorance is deliberate; it is built in to the philosophy of the blame culture — if they were to give sensible advice, they might be liable for the consequences!

This approach is unhealthy, and in general penalises those trying to do it right over those who "keep their heads below the parapet". Only by cooperation and education can this undesirable situation be reversed, and in part this book aims to allow enforcers and practitioners to at least start talking a common language!

Notes

1. See for explanation — http://onlinelibrary.wiley.com/doi/10.1002/9780470552940.app1/pdf
2. "Societal risk and the concept of risk aversion", J K Vrijling and P H A J M van Gelder, accessed via http://academics.ewi.tudelft.nl/live/binaries/ec906edd-af0e-4ba1-bcd8-18e7e7cd73ab/doc/paper10.pdf
3. See for example — http://www.hse.gov.uk/societalrisk/
4. See for critique — Cox, L A Jr, "What's Wrong with Risk Matrices?" Risk Analysis, Vol. 28, No. 2, 2008, DOI: 10.1111/j.1539-6924.2008.01030.
5. This section is adapted from a paper published by the author in the Journal of Pyrotechnics — see T Smith, Journal of Pyrotechnics, 29, 2010, 12–31 — http://www.jpyro.com/wp/?p=1164

6. UK risk guidelines — See http://www.hse.gov.uk/risk/. Note that explosive specific risk information will shortly be available on the UK HSE website
7. See http://www.hse.gov.uk/risk/theory/alarp.htm
8. See http://webcommunities.hse.gov.uk/connect.ti/explosives/ view? objectId=55760
9. MAGIC maps — See http://www.magic.gov.uk/website/magic

Chapter 6 – Planning the display site

Tom Smith - Davas Ltd

The last thing in the planning stage of an event is the layout and arrangement of the display site itself. By now all the pieces in the planning jigsaw are complete or almost complete:

- What the client wants
- Which fireworks will ideally be used
- Fallout under various conditions
- Low-frequency, high-consequence incidents

If the site is flexible then the choice of fireworks can be relatively flexible too, and final decisions can be made on the site during setting up. If the site is "fixed" (by which we mean both the overall area and the firing area) then the choice of fireworks may be more limited.

Ideally the audience should only be on one side of the display site, and preferably the site should be oriented so they are upwind of the firing area (so that debris falls away from them) — see Figure 6.1.

The fallout area is downwind and extends a significant distance (either calculated from the information or as a constraint on the types of material to be fired under specified conditions).

Of course not only is this often impossible because of other site constraints, but even when the site has been worked out with proper

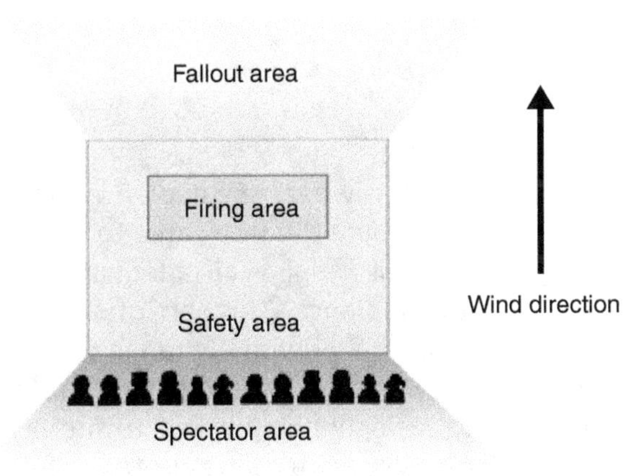

Figure 6.1 – Idealised site layout

consideration of the prevailing winds nature has a habit of reversing the wind direction during the setup and firing of the display!

More often the audience are on two or more sides of the display area and as a consequence the display may have to be much more "vertical" in design to prevent fireworks deliberately fired at an angle reaching the audience. Such a layout is shown in Figure 6.2.

The problem with this arrangement is when the wind at the firing time travels towards the audience. In some cases, and with accurate prediction, it is possible to move the firing area backwards in order to maximise the fallout area between firing point and the audience (Figure 6.3), but all too often either the wind change is not predictable or the site does not allow the firing area to be changed.

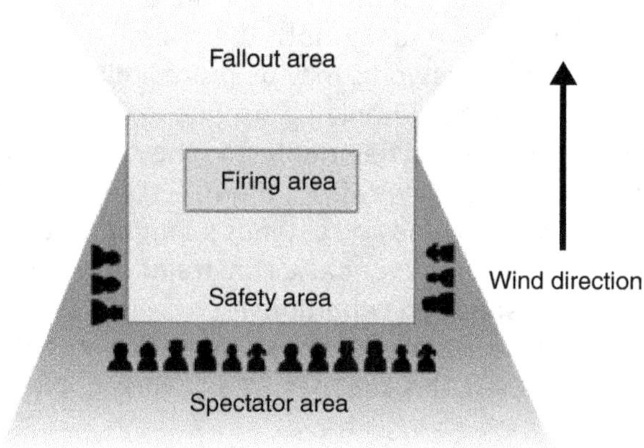

Figure 6.2 – Audience on 3 sides

The worst case, of course, is where the audience is on four sides and there is little room to manoeuvre the firing area to maximise the down-wind fallout range while still keeping an adequate "safety" distance from the fireworks to the audience. This sort of situation is, I believe, too complex for an "amateur" firer with the range of materials available except where ALL the distances are great. For the "professional", this is a situation they are increasingly being asked to fire in — particularly when firing from structures, or from rooftops etc.

In such cases the choice of fireworks is necessarily very pessimistic; almost every wind direction presents a "worst case", with debris

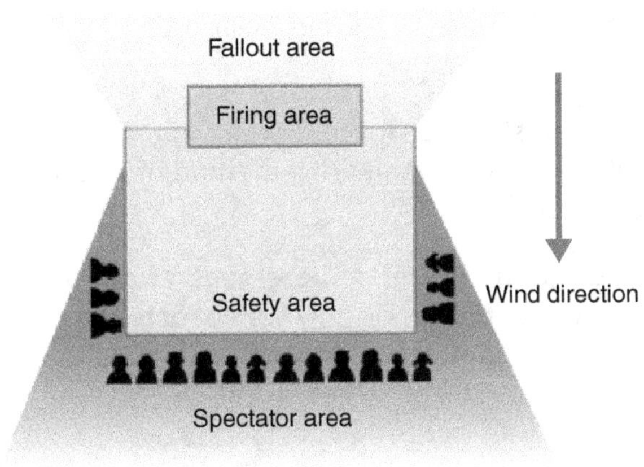

Figure 6.3 – Non-ideal arrangement — wind towards crowd

potentially falling on the audience. To some extent this problem is exacerbated by the elevated firing position (e.g., a rooftop) so that the debris has a longer time to fall to the ground, and can thus travel farther. In addition, such rooftop firing sites are rarely isolated, and the winds tend to be unpredictable and funnel between and around buildings.

As a consequence, the range of fireworks to be fired is very limited and usually consists of

- Single-shot comets
- Single-shot mines
- Fountains

Multi-shot Roman candles have wadding between the shots, and any thought of using shells or rockets should be dismissed.

Layout of the firing site

There are many ways to lay out the firing site, and obviously much depends on the nature of the site layout and the design of the display. In many cases the firing area will be entirely dictated by the nature of the display (e.g., from a structure) or the available space (e.g., from a barge).

The following section is included to illustrate some of the possibilities available and to highlight some of the pitfalls we have witnessed.

Overall the principles are, in no particular order:

- Ease of setting up
- Minimising the possibilities for mistakes
- Compatibility with the chosen firing method or system
- Safety of the firers

The following diagram illustrates these general principles, but should not be taken as the definitive or only way of achieving them. Alternatives, particularly in relation to mortar arrangements, are described later in this chapter.

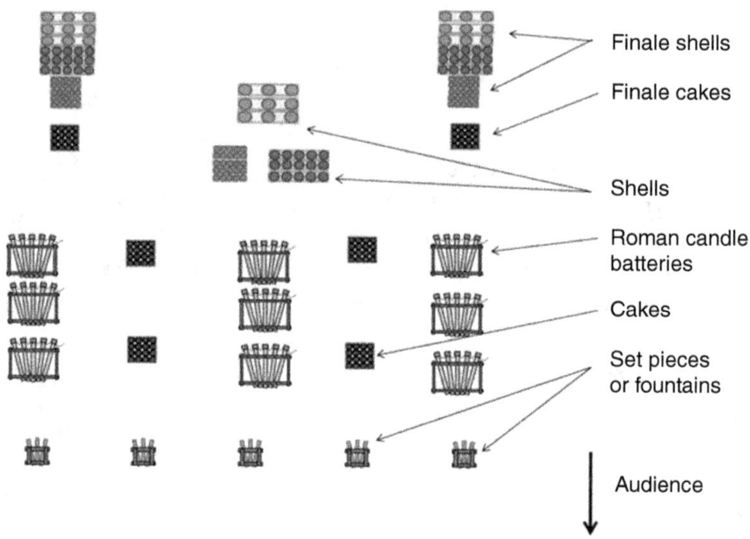

Figure 6.4 – Generalised site layout

The features of this generalised layout are as follows (Table 6.1):

Table 6.1 – Features of generalised site layout

Feature	Layout and comments
Towards the front of the site (nearest the audience)	Non-projectile effects, e.g., set-pieces and fountains. Lancework can also be places near the audience.
Towards the rear of the site (farthest away from the audience)	The aerial effects (shells and larger-calibre Roman candles)

Feature	Layout and comments
Similar types in rows	Facilitates setup and prevents matched pairs of items being misfired
Space between rows	Allow walkways for ease of setting up, for firing (if manual) and for cable runs (if electric)
Number of posts (firing positions)	There is a temptation to have too many different positions in a small/medium display. Occasional firing from multiple points gives variety, but usually there is no need to fire from too many points simultaneously — it just looks messy!
Finales	If possible keep upwind (but often this is impossible) and away from other fireworks. Cover to prevent accidental premature ignition.

Basic mortar layouts

There are a number of ways that the different firework types can be laid out within the firing area — and we will illustrate various possibilities using mortar racks as an example.

Obviously all that follows is constrained by the nature of the site, the area available, and the design of the display. It may, for instance, only be feasible to fire shells from a single position, or it may be that while this is true for the majority of the display, during the finale simultaneous firing from several positions is desired.

Sometimes the layout of the mortar racks (and indeed the whole site) is determined by the firing system used and the way it must be positioned to minimise cable runs and the possibilities for errors.

Three common types of mortar layout in frequent use are illustrated in Figure 6.5 below.

These are:

- By sequence
- By calibre
- Separated

Each has advantages and disadvantages, and these are outlined in Table 6.2.

Numbers of firing positions

The aesthetics of show design are discussed in Chapter 11, but the fundamental principles in terms of site layout are best discussed here.

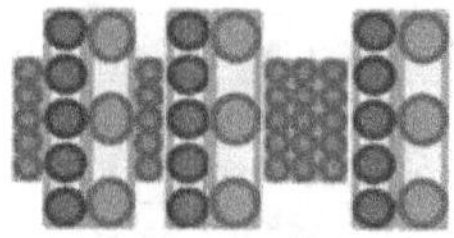

Arrangement A – in sequence

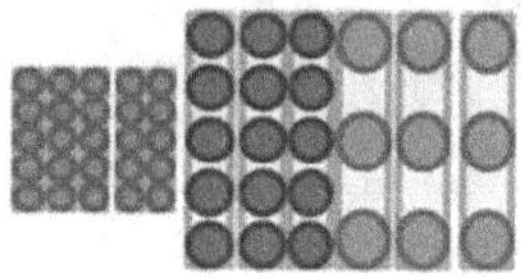

Arrangement B – by calibre

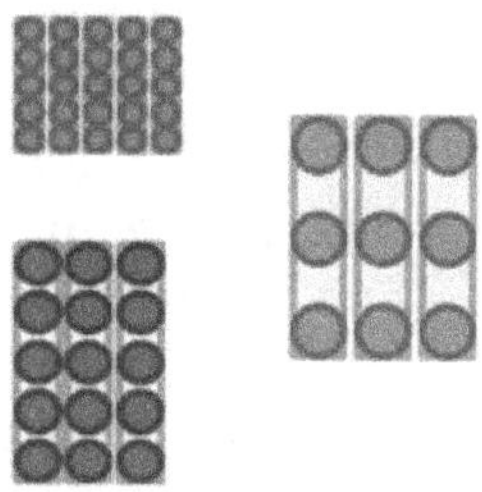

Arrangement C – separated

Figure 6.5 – Possible generic mortar layouts

Table 6.2 – Layout of mortars

Factor	**Layout by sequence**	**Layout by calibre**	**Separated layout**
Ease of setting up	Easy to do on-site	Requires pre-event planning	Easy to do on-site
Use of pyrotechnic delays within a sequence	Simple if single ignition per sequence is used. However, transporting sequences of mixed items may cause problems.	Requires either independent pyrotechnic delays (all ignited simultaneously by a single cue) or by sequential cues	Requires either independent pyrotechnic delays (all ignited simultaneously by a single cue) or by sequential cues

Factor	Layout by sequence	Layout by calibre	Separated layout
Minimising risks from mortar failure	Use of old-style crates leads to significant risks from mortar failure affecting adjacent unfired tubes. "Ladder" racking systems are not used in this configuration.	Use of old-style crates leads to significant risks from mortar failure affecting adjacent unfired tubes. However, this layout also lends itself to modern "ladder-type" racking systems.	Use of old-style crates leads to significant risks from mortar failure affecting adjacent unfired tubes. However, this layout also lends itself to modern "ladder-type" racking systems.
Placement on the display site	Sequences can be set up at a number of posts. It is rather doubtful that exact placement is necessary, except when "tailed" shells are fired in "V" type patterns — in general a small difference in the firing position is completely outweighed by variations in flight.	This arrangement can be less precise than by sequence.	This arrangement can be less precise than by sequence.
Wiring considerations	The simplest wiring arrangement — but often relies on pyrotechnic delays between the elements of a sequence. This may be acceptable for smaller events but rarely has the precision required for larger events or where simultaneous firing from several posts is required.	This arrangement can be used with multiple firing cues per sequence (e.g., from a modern firing system) — but introduces potential issues during wiring up — see later.	This arrangement can be used with multiple firing cues per sequence (e.g., from a modern firing system) — but introduces potential issues during wiring up — see later.

A modern site layout on an open site generally consists of a number of "posts" or firing positions. Fireworks that produce an effect from ground level upwards (e.g., mines, candles and fountains) are usually lined up on these posts so that the patterns produced are symmetrical and even. Shells and other items can be fired from similar positions, or often are laid out on the site between the "posts" as shown in Figure 6.4. It is important that the number of posts is appropriate for the site and the display design (even if Roman candles are fired from a maximum of five positions it may be appropriate to have 11 posts — each alternate post corresponding to a candle position, and the intermediate ones corresponding to other effects), measured accurately and clearly identified on the site plan and the display plan and layout. For the largest of displays, each post may be rigged separately and almost independent of the others (but see cautions on wiring below), with equipment and fireworks supplied specifically for that post. Such an approach relies on all the firers working with a common set of techniques and with a clear understanding of the importance of following the display plan exactly!

Wiring issues

Presuming the display is to be fired electrically, however the display site is laid out and whatever firing equipment is used there are a number of potential issues that must be considered — the most important of which, from an aesthetic point of view, is that the wires are connected correctly! This may seem a stupid and almost trivial thing to say — of course the display will be connected correctly — but far too often we have witnessed displays where an elementary error has been made. For instance, if firing single mines, or Roman candles from five posts, we have all seen the situation where there are four posts of silver and one red (or whatever). Sometimes this can be deliberate — but far too often we see the inevitable consequence later on in the display where four positions fire red and one fires silver!

The more complicated the site layout the more likely this is to occur. It is not a safety issue (in the main) — but it can ruin the overall effect. It may not be noticed by the vast majority of those watching — but for the highest-profile events where an iconic image is paramount, it can look very poor indeed!

The same problem arises with shells — indeed because there is less external distinction between types (once in a mortar all shells look the same) the problem can occur more often. There is no simple solution to this problem — except to instil in those doing the wiring the need for precision, and if necessary a need to have their work checked. Old-fashioned firing layouts, using vast lengths of bell-wire, were, perversely, less likely to have this problem when a rigger moved from post to post, laying a wire and connecting the firework fuses (usually attached to fireworks of the same type) together. Modern firing systems where individuals may wire all the fireworks to a splitter box at a single post — but not work at all the posts — have led to the problem becoming somewhat more common.

Repeated displays

Where a display is to be fired repeatedly from the same site (or essentially the same display is to be fired at least) there is justification for permanent or semi-permanent installation of some of the rigging equipment, firing lines etc. This can range from permanent mortar installations (as, for instance, at some competition sites like La Ronde in Montreal) or at a site for repeat performances on, say, consecutive days, posts arranged for rigging of candles and mortars and the multi-core cables for wiring.

Probably the ultimate approach to this is at the various Disney sites around the world — where permanent installations are common and the displays effectively fired are identical over an extended period of time.

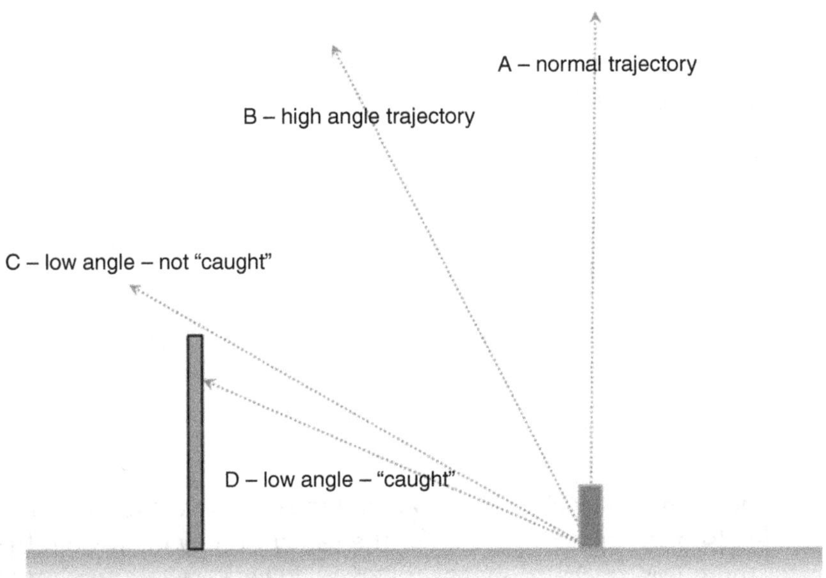

Figure 6.6 – "Catchers"

Catchers

An analysis of the risk-reduction methods that are viable at such sites may reveal that permanent safety installations may be justified — the Disney sites use “catchers” to reduce the risks to the audience and the site from low-trajectory shell firings.

It is critical to evaluate such devices carefully and not to assume that any barrier between the firing area and the audience will necessarily “catch” all low-angle discharges.

If a single “catcher” is used at the display site, then mortars farther away from the “catcher” are less protected than those nearer to the “catcher”.

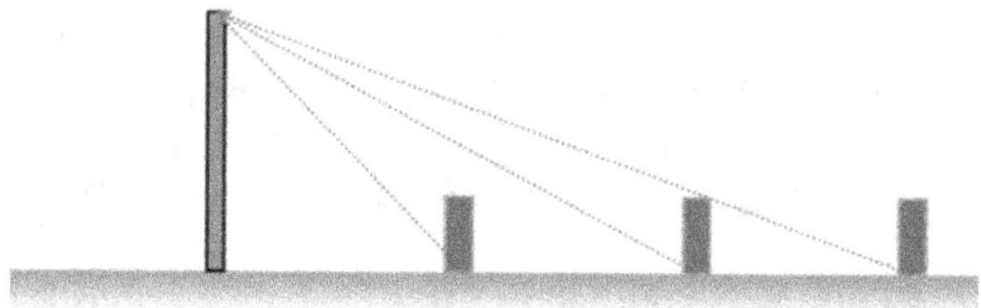

Figure 6.7 – Different catching angles dependent on the position of the mortars

In some circumstances the only way to overcome this limitation is to have multiple rows of “catchers” — one for each row of mortars.

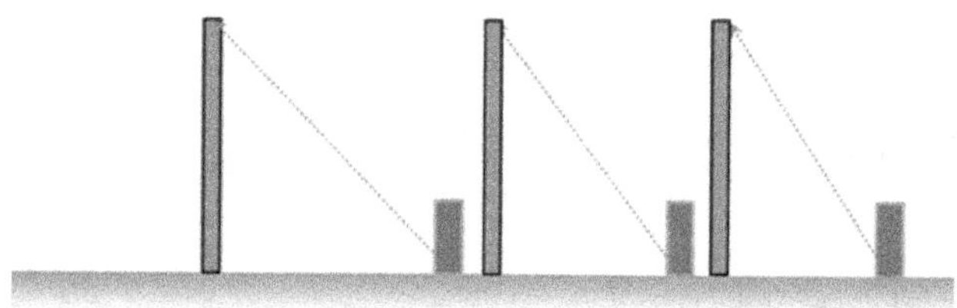

Figure 6.8 – Multiple catchers

It is important that the “catchers” are not too close to the mortar racks — otherwise the shells can puncture the catcher on ejection from the mortar tube. This inevitably means that the catchers are tall, and hence generally expensive. The analysis of costs and benefits outlined in Chapter 5 would rarely indicate that such protective methods are justified.

Chapter 7 – Fallout

Tom Smith – Davas Ltd

It is important to recognise that "what goes up must come down" — there is almost always some debris produced from fireworks, and this debris, even from normally functioning fireworks, can pose a significant hazard and therefore risk. In addition there may be "abnormal" fallout from a malfunction in the firework, such as:

- A shell delay fuse failing and the shell falling to the ground where it may subsequently burst
- A low-burst shell
- The failure of a support for a firework either leading to incorrect orientation when firing, or disruption of the support itself

The following diagrams (Figures 7.1–7.5) illustrate these different types of shell failures and the resulting potential for harm to operators, spectators or structures.

Both types of fallout and the risks associated with both types need to be considered in planning the display and producing an overall risk assessment. It is not sufficient to determine the "safety" distances on either abnormal functioning or "normal" fallout but not both. "Normal"

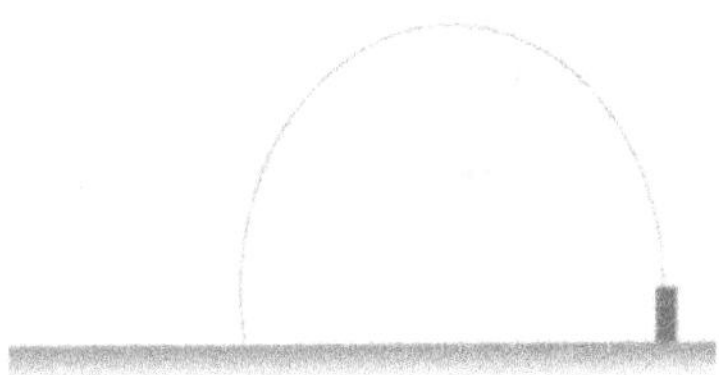

Figure 7.1 – Dud shell

Figure 7.2 – Low-burst shell

Figure 7.3 – Ground-burst shell

Figure 7.4 – Muzzle burst or very low burst

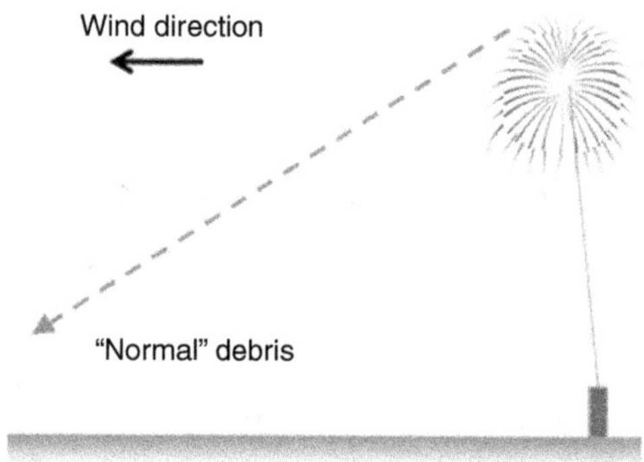

Figure 7.5 – "Normal" debris

fallout will occur for every firework (or shell) fired, although the extent and nature of this fallout will differ according to shell type and calibre and the potential for serious harm is low — this could be termed a "high-frequency/low-consequence" type event. The alternative low-frequency/high-consequence of a shell bursting over or in the crowd is of course much more serious, but the risks from the two modes may actually be similar. Remember that, in the simplest terms:

Risk = Frequency × Hazard

There is no simple way to relate a "safety" distance either to the shell calibre or the shell burst height. For instance, the French approach of calculating the minimum "safety" distance at 0.8 × the shell burst height is completely arbitrary — it could be argued that if the shell ascends and performs "normally", then the safety distance could be just a few metres. On the other hand if the mortar discharged at a low angle, the shell would travel much farther than 0.8 × apogee. Furthermore, a fixed multiplier related to calibre (as in the US) pays no attention to wind strength for "normal" debris fallout. All of this leads us to conclude that the only way to determine a distance at which the risk to the audience or other hazards is acceptably low (i.e., NOT a "safe" distance) is to perform the sort of calculations outlined in Chapter 5 using product- and equipment-specific data for each company and from which a meaningful set of performance and failure data can be developed.

Dangers from "normal" fallout

The following table (Table 7.1) illustrates the typical types of fallout that can be expected from a variety of firework types.

In the vast majority of cases identified above (with the exception of a rocket stick) the debris is not likely to cause fatalities, but major injuries can still occur from, for instance:

- Debris in an eye
- Lit debris causing burns
- Lit debris igniting clothes or other flammable materials

At the Melbourne Commonwealth Games in 2006 (see Chapter 20) an extensive study of the debris from the chosen fireworks was undertaken to determine the likely impact on

- The audience
- The athletes who were massed within the stadium
- The performers of the ceremonies

Taking into account the following factors

- Position of firing (the Melbourne Cricket Ground is almost circular and firing was undertaken from the entire circumference)
- The type of debris produced
- The weight, density and shape of debris produced

- Terminal velocity of debris produced
- Fallout patterns
- Population density of the various people "on the ground"
- The area they present to falling debris, together with area of face and eyes
- Effect on the eyes of debris of various shapes and energies impacting directly

the conclusions were as follows:

- A maximum of seven people could be hit by falling debris.
- None of these would have serious injuries from the debris.
- No lit debris would enter the audience or athletes' areas.

Table 7.1 – Fallout from firework types

Firework type	Potential "normal" fallout risks
Shells	Shell casings and components — post burst Parts of the lifting charge container Long-burning stars Unlit stars/units Component (e.g., whistles) cases
Roman candles	Bombette and component debris Long-burning stars Unlit stars/units
Rockets	The stick Head casings — post burst Motor (if separated from stick) Long-burning stars Unlit stars/units
Mines	Bombette and component debris
Fountains	Drossy debris
Set-pieces	Limited and local except if additional types are incorporated

The Melbourne Government approved this extensive risk assessment and quantification of the risks, and in fact the total number of people who were impacted by debris and who claimed was two — and a total claim of A$250. This extensive risk-assessment approach is justified on the largest of events where there are significant numbers of people at risk. We can only hope that the same pragmatic solutions and acceptance of risks are adopted widely elsewhere and that in time the approaches used are applied to smaller-scale events also.

Effects of wind

If we consider the effects of wind on "normal" debris the important things to consider are

- The height at which the debris is produced (which can be determined from measurements or calculation, or in the future for items in the European Union by the safety performance data supplied on the firework itself — see Chapters 3 and 13)
- The position in which this debris is produced relative to the firing point and the audience (i.e., if the shells are angled in whatever direction)
- Any initial velocity of the debris and its direction
- The nature of the debris (burning or inert, density, shape etc.)
- The wind direction and strength

From this it is possible to model the likely fallout pattern of this "normal" debris by making a few reasonable assumptions:

- That debris rapidly approaches terminal velocity downwards from the effect of gravity. Hence the time taken to fall to ground may be calculated.
- That the wind strength and direction remain constant throughout the fall of the debris (local effects may drastically change this — beware).
- Hence the lateral distance travelled by such debris may be calculated.

The following tables illustrate the calculated maximum fallout distances for "normal" shell debris (i.e., parts of the shell casing) of approximately 10mm × 10mm size. Smaller debris will travel farther. Similar calculations are possible for other fireworks, other calibres and specific effects (e.g., long-burning stars).

Table 7.2 – Normal fallout for shells fired vertically (distances in metres)

Shell calibre	Wind speed (Beaufort force)			
	Force 2	Force 3	Force 4	Force 5
75mm	41	69	97	126
100mm	49	83	116	150
125mm	56	94	132	171
150mm	61	104	146	188
200mm	72	121	170	220

Table 7.3 – Normal fallout for shells fired at 10° downwind (distances in metres)

Shell calibre	**Wind speed (Beaufort force)**			
	Force 2	**Force 3**	**Force 4**	**Force 5**
75mm	60	88	115	143
100mm	73	106	138	171
125mm	82	120	157	194
150mm	91	133	174	215
200mm	108	156	203	251

Table 7.4 – Normal fallout for shells fired 20° downwind (distances in metres)

Shell calibre	**Wind speed (Beaufort force)**			
	Force 2	**Force 3**	**Force 4**	**Force 5**
75mm	78	104	129	155
100mm	93	124	154	185
125mm	106	140	175	209
150mm	126	164	202	240
200mm	139	183	227	271

Table 7.5 – Wind speed comparison table

Beaufort force and name	**Effect on land**	**Wind speed**		
		ms^{-1}	**mph**	**knots**
0 – Calm	Tree leaves don't move, smoke rises vertically	0–0.2	<1	<1
1 – Light air	Tree leaves don't move, smoke drifts slowly	0.3–1.5	1–3	1–3
2 – Light breeze	Tree leaves rustle, flags wave slightly, wind felt on face	1.6–3.3	4–7	4–6
3 – Gentle breeze	Leaves and twigs in constant motion, small flags extended	3.4–5.4	8–11	7–10
4 – Moderate breeze	Small branches move, flags flap, raises dust and loose paper	5.5–7.9	13–18	11–16
5 – Fresh breeze	Small trees sway, flags flap and ripple	8.0–10.7	19–24	17–21
6 – Strong breeze	Large branches sway, flags beat and "pop", telephone wires whistle	10.8 –13.8	25–31	22–27

Beaufort force and name	Effect on land	Wind speed		
		ms^{-1}	mph	knots
8 – Gale	Twigs break off trees, wind generally impeded progress	17.2–20.7	39–46	34–40
9 – Strong gale	Branches break off trees, tiles blown from roofs	20.8–24.4	47–54	41–47
10 – Storm	Some trees blown down, damage to buildings	24.5–18.4	55–63	48–55
11 – Violent storm	Widespread damage to trees and buildings	28.5–32.6	64–74	56–63
12 – Hurricane	Severe and extensive damage	>32.6	>74	>63

Where fireworks are being fired at angles it is possible that the "normal" debris distance may be significantly extended in the direction of firing if the wind is also in that direction. The use of ShellCalc© (see below) to determine likely debris starting points is useful — the values of shell-burst displacement can be added to the wind-induced fallout calculations to obtain maximum values.

Table 7.5 is included to enable rough appreciation of the wind strengths. It is not expected that at very high wind strengths any fireworks would be fired — but the data are included for completeness.

Shellcalc©

Shellcalc© is a tool developed by John Harradine[1] of Queensland, Australia following investigation into an incident at Bray Park and modified by the author of this book.[2] Shellcalc© produces numerical and visual representations of shells and comets in flight for use in planning "safety" distances.

The program, written in Microsoft® Excel for convenience, has been extensively modified and allows inputs as shown in Table 7.6:

Table 7.6 – Shellcalc© input parameters

Parameter	Comments
Type of effect Shell or comet	If "Shell" is selected a burst diameter is calculated and a possible fallout trace is produced. If "Comet" is selected then the projectile is assumed to be consumed during flight.
Units	Metric or Imperial

Parameter	Comments
Calibre	Diameter of projectile
Mortar angle	Direction fired from vertical
Muzzle velocity	Entered if known, otherwise standard values used
Fuse delay	Entered if known, otherwise standard values used
Shell mass	Entered if known, otherwise standard values used
Tumbling/mortar drift	Allows a factor to be introduced to simulate barrelling and tumbling effects. If this is not present, a shell fired vertically would land exactly where fired if no wind parameters are entered.
Wind speed	At ground level
Relative wind direction	Relative to the firing angle
Elevation of launch site	To allow for changes in atmospheric density etc.
Height of launch	Height of launch point above the ground — introduced to allow for firing from buildings and to permit debris to extend below launch height
Terrain category	To accommodate open, wooded or built-up areas

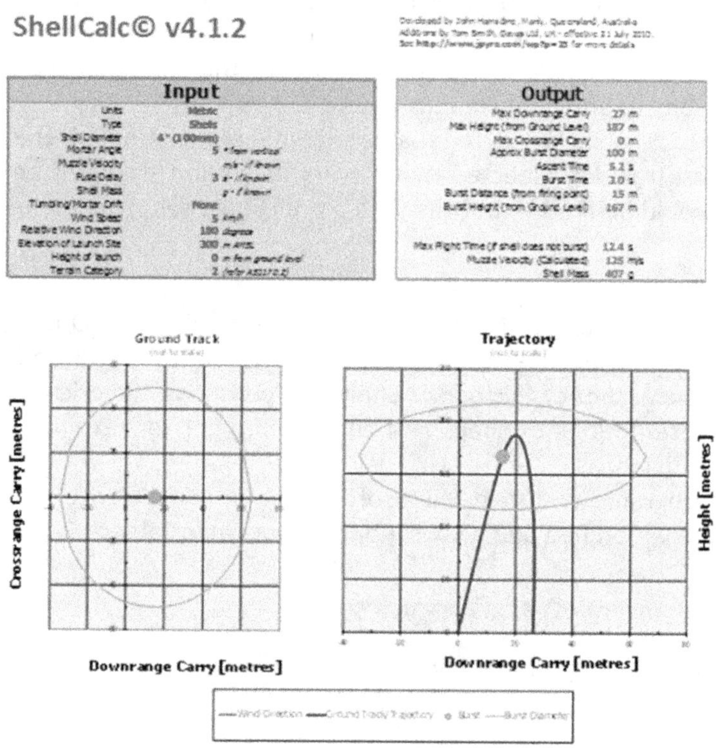

Figure 7.6 – 4" (100mm) shell — Fired 5 degrees from vertical with low wind

Output from the calculations is produced graphically and in a tabulated form. A selection of typical outputs is given below. Note that the depiction of the shell-burst diameter may be distorted depending on the axes that the program selects to display the results — the alternative would be a highly extended "y" axis in the majority of cases.

Shellcalc© can also be used to model the trajectories of comets, which is particularly useful for the risk assessment of firing single-shot items from buildings. In this case Shellcalc© allows modelling of below-horizontal firing and allows modelling of "blowback" onto the building in extreme conditions (Figure 7.8).

Taken together, the calculated data from "normal" debris distances (which are relatively low-hazard but high-frequency) and shell failures (which are high-hazard but low-frequency) allow the display designer to design a display that presents an acceptably low risk, while also providing objective criteria for curtailment or cancellation of the display.

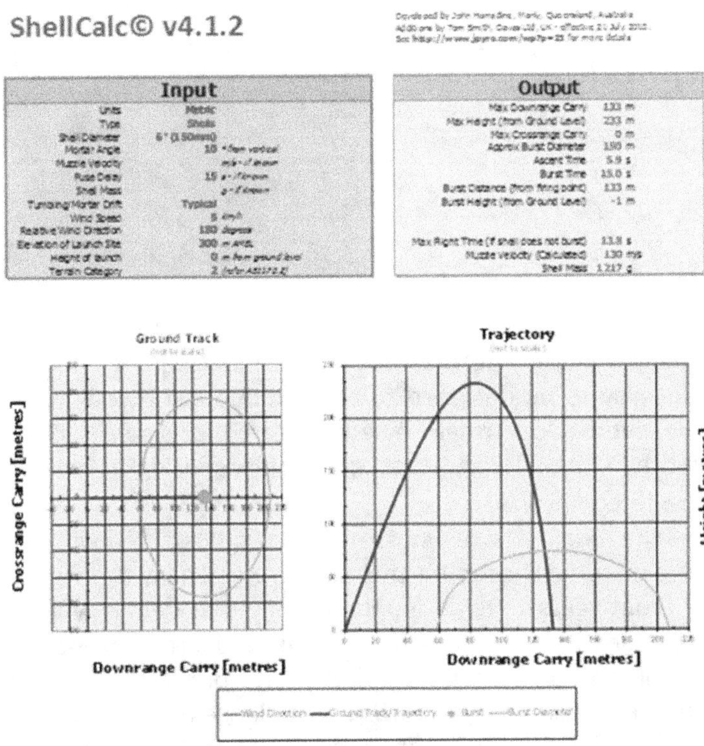

Figure 7.7 – 6" (150mm) shell with deliberately extended fuse time (i.e., shell fuse failure). Note the position and extent of the predicted burst.

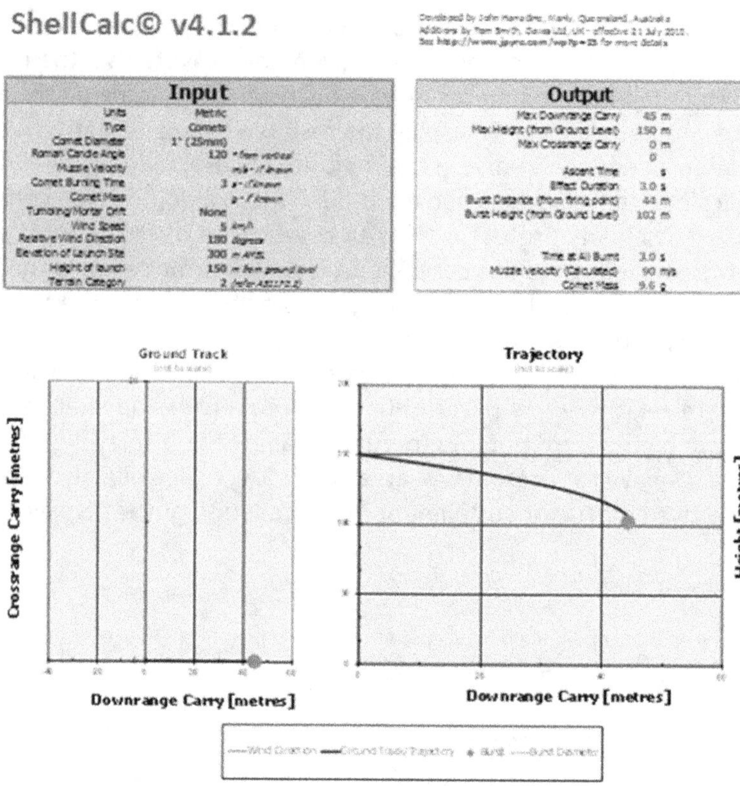

Figure 7.8 – Comet fired from a building

We usually work on a normal maximum wind speed of Force 4 — this allows a display to be designed to accommodate the vast majority of conditions that would normally be encountered for displays. If the wind is stronger than this, it is likely that other safety considerations will need to be considered anyway.

From all the data we have amassed and calculated, and with a knowledge of the material chosen for the display, we usually then prepare an objective "contingency" table for the display that details what curtailments will need to be made as a function of wind direction and strength. Direction is important because it may be that from certain directions an additional risk is posed (i.e., towards the audience or other hazards), but if the wind is blowing in the opposite direction then a greater wind strength can be tolerated before curtailment or cancellation is necessary.

Table 7.7 is an extract from one such table.

What is important, especially in large-scale events, is that:

- All parties involved in the display "sign up" to the contingencies — this includes the event producers and clients.
- The broadcast media are informed of the various contingencies and can plan alternative shots if the conditions dictate.
- There is a means of assessing the wind direction and strength.
- There are mechanisms for curtailing or stopping the display if debris falls in an unpredicted way (perhaps due to local wind funnelling) or if a failure occurs.
- There is adequate, dedicated communication between the firing site, the event producers, the advisors and "fire watchers" and the emergency services in case a decision has to be taken to curtail or cancel the display.

Table 7.7 – Extract from contingency table for London NYE display

Wind direction	Wind strength	Consequence	Action
N	<Force 4	Debris falls as predicted.	None
N	>Force 4	Debris area is extended upriver.	Remove 150mm shells. Remove long-burning stars.
N	>Force 6	Debris falls on Westminster Bridge and potentially Parliament.	Remove 150mm shells. Remove 125mm shells. Remove long-burning stars.
N	>Force 8	Debris falls on Westminster Bridge and Parliament.	Cancel river part of display.
NE	<Force 4	Debris falls as predicted.	None
NE	>Force 4	Debris area is extended upriver.	Remove 150mm shells. Remove 125mm shells. Remove long-burning stars.
NE	>Force 6	Debris falls on Westminster Bridge and Embankment.	Remove all shells.
NE	>Force 8	Debris falls on Westminster Bridge and Embankment	Cancel river part of display.

(*Note:* We tend to colour all our tables green, amber and red to reflect the seriousness of the contingency needed.)

Some of the decisions can be taken before the display on the basis of predicted information — for instance in London it is essential that if the decision is made to cancel the New Year's Eve celebrations it is done before noon so that people can be advised not to enter the city. After this time there must be some sort of display, even if it is severely curtailed, or fired from a single point.

Notes

1. J Harradine & T Smith, Journal of Pyrotechnics, 22, 2005, 9–15 — http://www.jpyro.com/wp/?p=23
2. See revised version of Shellcalc© at http://www.jpyro.com/wp/?p=1193

Chapter 8 – Rigging the display

Tom Smith – Davas Ltd

There is no point in trying to dictate a single rigging method for all the different firework types and all the different customs used throughout the world. There are almost as many ways of rigging a display as there are display companies.

Whatever the type, scale or nature of the firework display there are some fundamental principles of rigging that always apply:

- The equipment used to rig fireworks (frames, mortars, racks etc.) should be designed and manufactured in such a way as to minimise the possibility of failure.
- Fireworks should be rigged so that there is a minimum possibility of them discharging at an undesirable or unpredictable angle.
- There should be, wherever possible, a "failsafe" method of rigging, which means that should a failure of a particular piece of the equipment occur, there is the least possibility of the failure impacting on the safety of the display.
- In general, no firework that has a projectile effect should be erected in a manner with a potential single point of failure — for instance, Roman candles should be affixed to a frame with at least two independent fixings, and in turn the frame should be affixed to a post with at least two independent fixings.

These principles apply to displays of any size and to any design of new equipment or firing ancillaries. The word "minimise" is deliberate — it is often impossible, or financially impractical, to eliminate the risks, and it is hoped that other chapters in this book will demonstrate the principles behind proper risk assessment. However, one should never be complacent, or assume that there will never be a failure of a particular firework in a manner that could cause failure of the rigging.

Rigging methods

The following are general principles applied to various firework types and should not be taken as definitive. However, they are based on principles established over many years.

In the case of fireworks for consumers, the method should not be changed from that described on the firework label and accompanying documentation — failure to abide by the written instructions can often be the cause of incidents. We have been involved in many cases where, for instance, the labelling on a consumer item required it to be "inserted upright in soft ground" but instead the firework was attached to a post — in this case it is likely that a court or insurance company would determine that the instructions were not followed by the user, and the incident deemed a result of user error.

Digging in

If any firework is to be dug into the ground, or supported by earth (e.g., in a drum) or by sandbags, then the essential criteria are:

- The support must keep the firework in the desired orientation.
- It should not lead to a diminution of the structural integrity of the firing device (e.g., mortar) by allowing water ingress. It may be appropriate to bury the mortar in a plastic bag to prevent this.

The purpose of this type of support is not to contain the firework, or fragments of a mortar, in case of product failure — the complications of trying to achieve this are simply too great and too unpredictable.

Where external supports are used to ensure the correct orientation of the firework, they should be constructed and used in such a way that the firework cannot fall over in normal functioning. In addition, the frame should be placed, where possible, on the rear of the supports (relative to the audience or other hazards) so that if the fixings become loose the device would be failsafe.

For single-firing items the tying methods can be quite simple, since the firing of the shot cannot affect a subsequent firing (there is none). For fireworks that fire multiple shots it is essential that the supports maintain the correct orientation of firing throughout the firing. The fixings for Roman candles in particular should not become loose during firing, since once started it is impossible to cease the functioning of the firework and if it is mis-oriented, shots may be fired towards the crowd or other hazards.

There are a number of tying methods used to affix fireworks to frames and frames to supports (Table 8.1).

Table 8.1 – Tying methods

Method	Common use	Comments
Wire	Candles onto frames	Care must be taken to ensure the wire does not fail where it is twisted together.
Adhesive tape	Bundles of candles	Adhesive tape offers little resistance to fire and the structural integrity may be compromised.

Method	Common use	Comments
String, rope or straps	Attaching frames to support posts Attaching single mortars to supports	Knots that cannot become loose should be used. Non-flammable (or non-melting) rope should be used.
Cable ties	Attaching candles to frames Attaching frames to supports	Cable ties of sufficient dimensions should be used — cable ties can be combined to produce longer dimensions.

Figure 8.1 – Simple Roman candles on frames affixed by cable ties. Note independent fixings (see text)

Rigging on solid surfaces

Where it is not possible to fix the frame to a support driven into the ground, it may be possible to fix the frame to an existing support, or to a specially designed support (see Figure 8.4) or to make an effectively self-supporting structure from many frames of the same dimensions joined together with battens etc.

Whatever method is chosen, the same general principles apply. We repeat them here for clarity:

- The firework should be fixed in such a way that it does not fall over during functioning.
- It should be fixed by at least two independent fixings.
- Where possible, it should be fixed in such a way that if failure of the support or fixings occurs, it is failsafe.

Figure 8.2 – 100mm mines — with line numbers and two independent fixings

The same principles apply to single-shot devices affixed to a support to produce a "chase" or other spread of effects.

Mortars for shells

The correct rigging of shells is of particular concern for many reasons:

- They are the most powerful projected items and hence are usually the determinant of overall "safety" distance.
- Large displays use a large number of shells of varying calibres and effects.
- Shells are often fired from mortars within racks — so that the failure of a single shell can affect the subsequent firing of shells in adjacent mortars.

Many types of mortars have been used. In many cases mortar tubes have not been specifically designed but are adapted from other,

Figure 8.3 – Roman candle batteries in which the individual candles are affixed by wire to the frame, and the frame attached to the support posts by cable ties

Figure 8.4 – Roman candle batteries supported by metal frames. Photo: A DiPalma

Figure 8.5 – Single-shot comet and mine sequence. Photo: A DiPalma

Figure 8.6 – Fanned single-shot comets. Photo: A DiPalma

suitable-diameter tubes that are available commercially. However, not all types of tubes are suitable for firing shells, and tests should be carried out (both in normal functioning and to simulate a shell bursting within a tube) to ensure that the mortar is suitable.

CANMET[1] has carried out extensive research on the catastrophic failures of mortars of various materials, and the research should be examined by anyone who is firing shells to see the likely effects should a mortar fail!

Table 8.2 – Mortar types

Type	Comments
Steel	Heavy Can fragment to produce sharp high-energy fragments
Iron	Heavy Can become brittle (from CO absorption) Can fragment to produce sharp high-energy fragments
Aluminium	Lightweight Can fragment to produce sharp high-energy fragments (although they have less range than iron/steel)
High-density polyethylene (HDPE)	Readily available — but not always exactly the right dimensions Custom-made end closures are available.
Glass reinforced plastic (GRP or FRP)	Lightweight Custom-made end closures are available.
ABS etc.	ABS tubes and other readily available plastics are often NOT suitable as mortars as they are brittle and fragment easily.
Convolute wound paper	Susceptible to damp May lose integrity Fragmentation hazards are less than metal mortars.
Spiral wound paper	Cheaper to produce than convolute wound Less long life than convolute wound Susceptible to damp May lose integrity Fragmentation hazards are less than metal mortars.

Figure 8.7 – Old-style mortar racks with HDPE mortars. Photo: A DiPalma

Mortar racks

Mortars for shells (and to an extent mines) are often loaded into racks for ease of transport and use on the display site.

Following an incident in the UK, Kimbolton Fireworks and the Health and Safety Laboratories (HSL) carried out extensive research (detailed in the following chapter) to determine optimum mortar rack construction as a result of shell failures.

The incident that initiated the research involved the bursting of a single spherical shell within a mortar, which caused the mortar to fail (i.e., fragment) and then caused an adjacent mortar to topple over before it subsequently fired. Once the mortar fired, the shell was projected towards the audience and exploded near to the crowd, causing significant injuries to one child and minor injuries to many more.

Two areas of research were followed:

- Means of containing mortars that had fallen over
- Means of preventing the failure of one mortar affecting adjacent mortars in a rack

The results of the first area were inconclusive but demonstrated that any normal barrier close to the mortar racks was inefficient (see information regarding "catchers" in Chapter 6, but the second yielded significant and reproducible results. The details of this research are given in the following chapter.

In racks containing mortars for shells over 100mm (4"), where the mortars were not separated from one another the deliberate bursting of a shell within a mortar led to disruption of the mortar rack and displacement of adjacent tubes. Separating the mortar tubes in racks for shells over 100mm (4") by a distance equal to the nominal shell calibre led to no disruption of adjacent tubes by deliberate shell failure.

From this research was developed a range of open-style racks (for shells of 100mm [4"] and below) and a style of "ladder" racks for shells over this size. Although at first sight such racks may appear inefficient and cumbersome to transport, by clever construction and efficient loading in a vehicle they have now been adopted in the shown or similar form by a wide variety of companies.

The racks illustrated in Figures 8.8, 8.9 and 8.10 incorporate the "ladder" part and are supported on open ground by metal poles driven into the ground. On solid surfaces the extent and weight of the rack make it effectively self-supporting.

Further developments could be driven by manufacturers, and we envisage the possibility of a set of racks that are self-stacking and made widely available at an economic price.

Similar designs of open-style racks have also been developed by other display companies. The use of such racks leads, in some circumstances, to revision of the site layout (see Chapter 6). The construction and extent of the racks dictate, to an extent, their arrangement on the display site.

Figure 8.8 – New-style mortar racks for 100mm mortars with spacing

Figure 8.9 – Small-calibre mortars in rack

Figure 8.10 – 125mm Mortars in "ladder" rack

Figure 8.11 – Close-up of 150mm "ladder" rack

Figure 8.12 shows the use of such racks with a simple tape covering to each tube — in this way it can easily be determined that each mortar has fired its shell at the de-rigging phase.

A similar racking system can be used for placement of finale sequences. In Figure 8.13 the first two racks are of the old style, as they are to be fired together. Hence the potential problem of a mortar failure affecting adjacent tubes in the same rack is eliminated.

Figure 8.12 – Mortar racks with taped tops to check firing. Photo: A DiPalma

Figure 8.13 – Finale mortar racks. Photo: A DiPalma

In some parts of the world hand firing and reloading of shells is still commonplace. Often this is because of the nature of the shells themselves (e.g., multi-break) and hence the demands on the mortars. Figure 8.14 below illustrates the still typical firing of large-calibre multi-break shells from single mortars supported in drums of sand in Malta.

Figure 8.14 – Hand loading and firing of a Maltese 6" multi-break shell. Note particularly metal mortars in drums of sand and position of helper straddling mortar prior to loading. Things have moved on . . .

Waterproofing

Unfortunately many displays are fired at times of the year when it can reasonably be expected to have inclement weather. In the UK, November must be one of the worst months to have a bonfire and fireworks!

Fireworks and firework compositions are very susceptible to moisture, and adequate means must be taken to protect them from moisture ingress. For example, if the lifting charge of a shell is damp, the shell may not be ejected from its mortar properly and may therefore fail to burst at the design height.

Modern construction can build in a high degree of waterproofing to firework articles at the production phase. Although the use of plastic components is discouraged (they do not degrade and they produce sharp debris) it is possible to waterproof paper and card components and to design closures that prevent water ingress. Failing this, it is possible to waterproof many fireworks prior to rigging by sleeving them in plastic, or containing them within a suitable plastic bag, or by treating the final article with waterproofing spray.

At the display site the final stage of waterproofing can take place, but care must be taken not to affect the performance of the fireworks in question. Wheels and moving devices in general should not be covered in plastic unless it can be easily removed prior to firing — inevitably if it is not easy it will not be removed and a wheel could then "strangle" itself and fail to perform correctly. Similarly, rockets will not fly correctly if covered with plastic bags!

In the case of projected items (shells, mines and Roman candles) it is possible to cover the mortars (or, in the case of Roman candles the

Figure 8.15 – Waterproofing of mortars for the 1997 Hong Kong Handover display

Figure 8.16 – Hong Kong crew 1997

individual tubes) with lightweight plastic and to fire through it without any noticeable deterioration in the projected height. In practice, for shells and bag mines loaded into open mortars, a combination of aluminised foil and lightweight plastic is often used to provide a spark-proof and waterproof covering.

Notes

1. See the Proceedings of the International Symposium on Fireworks, Montreal 1992 — contents page at http://www.isfireworks.com/Proceedings/Montreal%20proceedings.pdf

Chapter 9 – A study of shell failures and mortar rack construction methods

Darryl Fleming – Kimbolton Fireworks

The firing of aerial star shells from mortar tubes is considered to be one of the most hazardous operations for both operators and members of the public alike during firework displays that contain or use such items as part of the display. In the event of a malfunction the hazard can be significant, resulting, in a worst-case scenario, in multiple fatalities.

Kimbolton Fireworks have undertaken a significant amount of research and testing to ensure, as far as is reasonably practical, that aerial star shells cannot affect the audience even during a malfunction.

A shell can occasionally burst inside a mortar tube, and given the common practice of racking mortar tubes, there is a probability of an adjacent tube being affected.

Good practice at all firework displays where shells are being used should include a visual check of the mortar shell before loading, with an eye for any damage or compromised integrity that could lead to a malfunction. Additionally, it should be ensured that the shells are correctly loaded in the right-size tube; a shell loaded in an incorrect tube can lead to a low burst. It is unlikely that a shell can be forced into a tube that is too small; however, a shell can easily be loaded into a tube that is too big. It is advisable to have mortar tubes colour-coded according to size and diameter and racked in size order to reduce the risk of this occurring.

In the event of a shell being loaded in an oversized mortar tube, the gas from the lifting charge will not be sufficient to get the shell to its correct height; moreover the shell will be ejected from the tube and land back on the ground adjacent to where it was fired from. All mortar shells should be positioned at a sufficiently safe distance so that if a shell bursts on the ground near the firing position, the stars and projectiles cannot reach the audience.

The operators should undergo extensive training and achieve a considerable amount of experience in handling and loading shells correctly and indeed in firing them, especially if hand firing. The correct Personal Protective Equipment (PPE) should be worn, and an operator should be able to detect when a lift does not sound right or if a shell has malfunctioned. They should evacuate the area immediately and not take any unnecessary risks.

It is important to check that fuses are protected from getting damp, for instance from rain, by covering the racks with plastic sheeting. In addition, most fuses should be waterproof and the shells should be encased in waterproof plastic bags. While loading in wet weather use temporary structures or pop tents to work under, thus keeping the firework fuses and mortar tubes from exposure to the elements.

The mortar tubes themselves should also be checked thoroughly for damage and cracks and to ensure they are dry before being rigged and shells loaded into them. All structures that hold the tubes upright should also checked for stability and integrity.

There are a number of ways a shell can malfunction:

- The fuse does not light and the shell remains in the tube; this is not a significant hazard to the public.
- The shell is ignited and ejected from the tube, but the internal delay does not light; the shell then fails to ignite in the sky and comes back down (blind shell). This is of no significant risk to the audience; the safety distances are such that only the operators may be at risk. In general, the operators wear Personal Protective Equipment. In a hand-fired display the shell firer will be aware of a blind shell and will call a warning to the other operators to get out of the downwind location of the mortars. In larger displays fired by computer, no operators will be located in the fallout area.
- The shell ignites and only reaches a reduced height before landing back on the ground. This is usually the result of two reasons: the lift has been compromised and there is insufficient lift to eject the shell to full height, or the shell has been loaded in an oversized mortar tube. In either case the risk to the audience is generally low due to the safety distances imposed. The shell is unlikely to fall back to the ground any farther than 10 to 15m from where it was fired, and it will then burst on the ground, sending a blast of stars radiating out in all directions. Our tests have shown that a shell up to 150mm diameter bursting at ground level will not throw stars more than 50m from the point at which it lands. By adding on the 15m distance the shell may have landed from the point it was fired, the minimum safety distance the audience would need to be is 65m. Given that the minimum safety distance for all shells is 75 m, it is unlikely that stars will reach the audience.
- The shell explodes in the mortar tube. This is the most significant risk, and is addressed below. This again can be due to one of a number of reasons. First, the lift has deteriorated or fizzled, resulting in the shell functioning normally but remaining at the base of the tube to explode

at the predetermined time dictated by the internal time delay. Second, the shell casing or construction is faulty, allowing the flame from the lift to bypass the internal delay and reaching the burst and contents of the shell before it has left the tube. Third, the shell has been loaded into an excessively oversized tube; the shell functions normally but the lift is insufficient to allow the shell to leave the tube. This eventuality however will probably not result in a catastrophic failure, as the tube will probably remain intact and the stars will simply be ejected out of the tube upwards, resulting in a mine-like effect. The main hazard of a shell malfunction in a correctly sized tube where the shell explodes inside the tube is catastrophic failure of the mortar tube and the displacement of adjacent tubes, especially if fired from a normal rack.

The fuse may fail for one of a number of reasons, e.g., the igniter failed or became detached from the fuse, the fuse itself is incomplete or the fuse has become wet and fails to function. Regardless of how a shell fails to ignite, this failure will result in the shell remaining in the tube. Live shells are situated in all the tubes before and during the display until the display is over, and in theory all shells should have been fired. A shell failing to fire does not pose any greater risk than at the start of the display; in fact a failed shell is less likely to initiate at all, and the shell will simply remain in the tube until it is removed by the operators at the end of the display.

As discussed above, Kimbolton Fireworks do have a policy for checking for misfires and their removal. A period of ten minutes should elapse before operators approach the mortar racks to check for misfires.

An initial visual check will be made from a small distance of a couple of meters using a torch to sweep across the mortar tubes. The operator will also check for any sparks or smouldering paper that could provide a source of ignition. After this initial check, each individual tube is then inspected by the shell firer or senior operator; this is done with the use of an electric torch and a visual inspection down each tube. The operator is able to just see the base of the tube from an acute angle without physically allowing any part of their body over the tube. The shell will be removed and placed in a carton and sealed to be taken back to the factory. If a shell fuse has been initiated but has not continued to the lift, either due to being damp or faulty, there is a risk that the paper part of the fuse may smoulder down and the fuse suddenly continue to operate as normal, even after ten minutes have elapsed. In this instance the operator should pour water into the mortar tube to ensure the shell and fuse are not smouldering before the misfire is removed.

For additional supporting information relating to a blind shell, it is worth noting that Kimbolton Fireworks have a strict policy for rigging mortar tubes and racks to ensure that blind shells are not a hazard to the audience. All mortar tubes are rigged in the upright position downwind of the audience. Using Shellcalc© it can be seen that a 150mm shell with a 15 km/h wind blowing away from the audience will carry that shell approximately 69m downwind, well within the 100m fallout area. Kimbolton Fireworks require a minimum of 100m fallout area for 150mm shells and at least 100m safety distance from the audience. In some cases mortar tubes may be angled away from the audience to compensate for wind blowing towards the audience. The tube is never angled more than 10 degrees to the vertical away. In this situation, using Shellcalc© it can be calculated that a 150mm shell with a wind speed of 20 km/h will travel no farther than 82m from the point where it was fired, still within the fallout area. In extreme wind conditions, certain size shells will be removed from the display.

The research and testing undertaken by Kimbolton Fireworks have demonstrated that by employing separation distances between each mortar tube the risk of damaging an adjacent tube is very low, and the propagation from one tube to another is significantly reduced by introducing a separation distance. Kimbolton Fireworks use the diameter of the shell plus two inches as the given separation distance; therefore a using six-inch shell will require eight inches between tubes. The risk of an adjacent shell being ignited in the event of a failure is removed, and we thus feel that there is no need to introduce any other safety measures like sandbags or waterbags.

The method of rigging mortar tubes has been tested to destruction to establish the effect on adjacent tubes when a shell is intentionally ignited in a tube. Photographic and video evidence can be given upon request with sufficient notice. During our tests, shells were intentionally burst in the tube to establish what danger may exist. The tests demonstrated that in the event of shell failure inside a mortar tube, the tube involved was affected with debris travelling up to 25m. In all tests the adjacent tube was not drastically affected due to the separation distances; i.e., some minor scoring and marking occurred to the exterior of the adjacent tube, but the integrity of the adjacent tube was not breached and the live shell contained in it was not ignited or initiated. More importantly, the rack holding the tubes and all adjacent tubes was not damaged, and all other tubes in the rack remained upright.

The racking method is designed so as to create the minimum amount of confinement and therefore allow maximum escape of gases and reduce the forces involved. From the testing undertaken on the racks we are satisfied that the risk of an adjacent tube being knocked over is very low. Even if a tube did fall over, the separation distance between tubes means that it is even more unlikely that the shell contained in the disrupted tube would be ignited.

We believe that our method of racking has significantly reduced the risk of any dangerous occurrence in the event of a shell bursting in the tube. For computer-fired displays, where shells can be fired in quick succession and where the shells are located in a confined area like a barge, then a further control method is introduced: corrugated Herras™ fencing, which surrounds the shell racks. A disrupted mortar tube would fire the shell into the fencing. The shell breaks on impact and the stars contained are ignited, and no burst projectile hazard exists. If a shell should go over the fencing, the angle of the tube is sufficient that the shell would gain enough height and burst in the sky, posing no significant risk to anyone. It is important to note that our tests demonstrated that the significant hazard of adjacent mortar tube displacement was only likely to happen to shell sizes of 125mm plus. Our tests on 75mm and 100mm shells demonstrated that the forces involved during a failure did not result in breaching the mortar tube; as in this instance, the stars were ejected out of the tube in a mine effect and the rack and adjacent mortar tubes were unaffected. This is not the case with shells larger than 125mm, which require the safety measure mentioned above.

Figure 9.1 – Layout of "ladder" rack mortar crates on a barge

Background

The research undertaken by Kimbolton Fireworks came as a result of an incident at one of our November 5th firework displays in 2005. A large display synchronised to music and fired using FireOne™ was 12 minutes into an 18-minute program when what we believe to be a San Tai 150mm Happy Face shell burst in the tube. Although the burst communicated to two adjacent mortar tubes, the shells contained in these tubes ignited but performed normally, almost reaching their full height. A third mortar tube was affected; the shell in this tube burst at ground level. It was at this point that we believe the racking system holding the remaining tubes was disrupted. The result was a 150mm shell being ejected at a significantly reduced angle towards the audience.

The audience was approximately 175m from the mortar position, and the 150mm shell stopped short of the audience line when it burst, injuring 11 people, one of whom received serious injuries to the lower part of the leg.

The probability and frequency of this event are incredibly low; however, Kimbolton Fireworks initiated a program of research to come up with a suitable racking method that would not lead to disruption in the event of a premature failure in the mortar tube. The various methods of racking were tested to destruction to evaluate the merits of each construction method and to establish the extent of destruction and the likelihood of adjacent tubes being disrupted.

Shell burst patterns

The first test involved the ignition of various shells of different sizes on the ground to establish the distance the stars travelled. Our open airfield testing site was used, which offers 750m safety distance to the nearest road or pathway.

The shell was placed loose on the ground, with an igniter inserted into the fuse leader and ignited remotely.

We tested a wide range of shell types, from 75mm up to 200mm; each shell size tested included a selection of various types of effects from the following manufacturers: An Ping, Forward, Haga Premium Brand, Igual, Kimbolton, Lidu, Panzera, Pirofantasia, San Tai, Shogun, Soldi, Sunny, Vulcan and Zink.

The effects tested ranged from chrysanthemum and peony shell bursts to pattern effects, palm shells and polyp shells containing large comets.

In order to establish the distance the stars travelled we placed $1m^2$ white sheets of plywood at 10m intervals in both directions from the position the shell was placed. The results were filmed at dusk, and from viewing the playback of the tests we could establish the distance the stars travelled at ground level using the 10m interval marker boards.

The results varied according to size and type. The correlation of distances is shown in Figure 9.2.

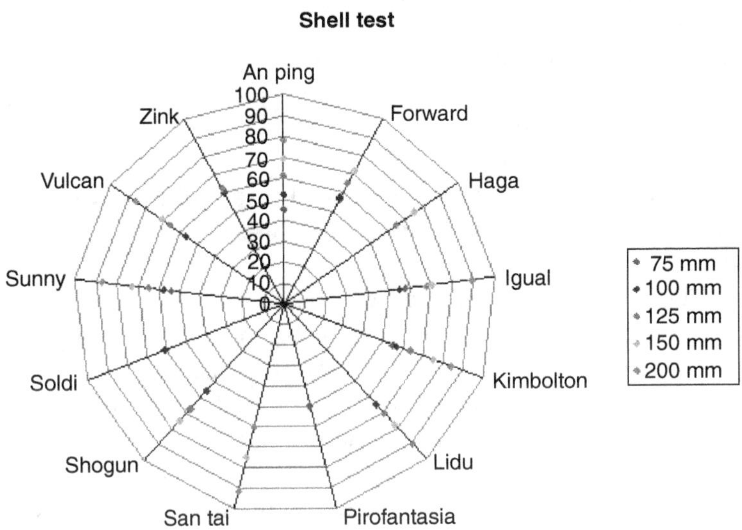

Figure 9.2 – Distance stars travelled at ground level for a variety of shell types and calibres

Shell failures inside a mortar tube

The next tests conducted were used to establish suitability of various construction methods of mortar tubes and to assess the likelihood and frequency of a premature shell failure in a mortar tube communicating to an adjacent mortar.

Shell failure was achieved by testing a single detonation of a shell while still in the mortar tube. The test was conducted by firing a single shell loaded in the mortar tube with the lifting charge removed. The shell was ignited remotely.

Shell sizes tested ranged from 75mm to 150mm and the mortar tubes were attached to an upright wooden stake at our airfield test ground. A second tube was attached to another wooden stake and positioned flush against the test mortar tube. A live shell was loaded into the second tube to establish whether any breach of the adjacent tube occurred and whether such a breach would lead to the ignition of the adjacent shell.

Various shell types and sizes from all manufacturers stocked by Kimbolton Fireworks were tested. The test also included various mortar tubes, including paper or card, HDPE and fibreglass. Steel mortar tubes were discounted due to the widely held view that they are not a suitable construction method for mortar tubes because of possible fragmentation.

We observed the following results:

Paper or card mortar tubes

- No 75mm shell tested breached the mortar tube (flower-potted). Maroon shells were tested and a full breach of the mortar tube resulted.
- Most 100mm shells tested did not breach the mortar tube (flower-potted). Some shell types did breach the mortar tube, in particular multi-break report shells from Soldi and some Lidu shells.
- Most 125mm shells resulted in a mortar tube breach but not all. Not all the failures resulted in a complete destructive breach.
- Most 150mm shells failed, resulting in a full breach of the mortar tube, and in some cases a full destructive breach was recorded.
- Card mortar tubes were found to be most likely to breach in premature detonation.

HDPE mortar tubes

- No 75mm shell tested breached the mortar tube (flower-potted). Maroon shells were tested and a full breach of the mortar tube resulted.
- Most 100mm shells tested did not breach the mortar tube (flower-potted). Some shell types did breach the mortar tube, in particular multi-break report shells from Soldi and some Lidu shells.
- Most 125mm shells resulted in a mortar tube breach but not all. Not all the failures resulted in a complete destructive breach.
- Most 150mm shells failed, resulting in a full breach of the mortar tube, in some cases a full destructive breach.
- HDPE tubes were regarded as the strongest construction; however, breaches were recorded by some shell types at 100mm, 125mm and 150mm.

Fibreglass mortar tubes

- No 75mm shell tested breached the mortar tube (flower-potted). Maroon shells were tested and a full breach of the mortar tube resulted.
- Most 100mm shells tested did not breach the mortar tube (flower-potted). Some shell types did breach the mortar tube, in particular multi-break report shells from Soldi and some Lidu shells.
- Most 125mm and 150mm shells resulted in a mortar tube breach but not all. Not all the failures resulted in a complete destructive breach.

Recommendations & conclusions

The following conclusions were made from these tests.

- 75mm shells fired from fibreglass or HDPE mortar tubes did not result in a breach of the tube.
- 75mm maroons did breach all types of mortar tube to varying degrees. It is Kimbolton Fireworks' policy to bury all mortar tubes containing maroons of any size in the ground or in steel drums filled with sand.
- 100mm round shells did not result in a significant breach of the mortar tubes made from fibreglass and HDPE. It is Kimbolton Fireworks' policy to fire any cylindrical shell or any shell with report effects from HDPE mortar tubes.
- Due to the fact that 125mm and 150mm shells recorded a higher rate of mortar tube breach, it was concluded that further tests should be conducted to establish the effects of these breaches on adjacent tubes and more importantly the racks that hold them.
- Card mortar tubes were regarded as the least reliable and resulted in the most destructive breaches. Kimbolton Fireworks are phasing out card mortar tubes.

Separation distances between tubes

After further tests we established that prematurely detonating 125mm and 150mm shells posed a risk to adjacent tubes and the rack construction. We conducted further tests to establish what separation distances should be incorporated to prevent a shell failure in a mortar tube from igniting a shell loaded in an adjacent mortar tube.

This test was conducted using our oldest fibreglass mortar tubes. The donor shell was ignited and detonated in the base of the tube

Figure 9.3 – Setup of individual tubes for testing

by removing the lifting charge (as opposed to putting the shell in upside down). The tests were conducted on both 125mm and 150mm shell sizes of varying types and effects.

The tests involved three mortar tubes, each secured to a wooden stake. The donor shell was ignited in the centre mortar tube. The two tubes adjacent on either side were loaded with live shells of normal construction, with the leader fuse coming up to the top of the mortar tube.

The first test was conducted by placing the two test mortar tubes up against the donor tube with no separation distance. The test was then repeated a number of times by increasing the distance between each tube by 25mm each time.

Observation

The following observations were made from these tests:

- 125mm shells did not always breach the mortar tube, and the effects simply flower-potted.
- When a breach did occur occasional communication to the adjacent tube was recorded. A separation distance of 100mm reduced the frequency of communication to the adjacent tube.
- Most 150mm shells breached the mortar tube.
- Some breaches with 150mm shells resulted in communication to the adjacent shell when there was no separation. A separation distance of 150mm significantly reduced the frequency of communication to the adjacent shell.

Recommendations & conclusions

- A separation distance between 75mm shells and 100mm shells was not a requirement; however, it would be advantageous. Kimbolton Fireworks have a policy of introducing a separation distance equal to half the shell diameter in all our 75mm and 100mm racks, especially due to the high frequency of firing these sizes of shells in bouquets matched in pairs or threes.
- 125mm shells rarely resulted in communication to the adjacent tube; however, the frequency is high enough to require some sort of separation. Kimbolton Fireworks' policy has been to introduce a separation distance of 175mm between tubes.
- 150mm shells did result in communication to the adjacent tube where no separation existed, but not in all cases. Kimbolton Fireworks' policy has been to introduce a separation distance of 200mm between 150mm tubes.
- The separation distance between tubes achieves two results. First, the likelihood of adjacent tube being disrupted is significantly reduced. Second, the likelihood of communication to an adjacent tube is significantly reduced.
- While the frequency of communication was high, it is important to note that the frequency of a shell detonating in the tube is very rare.

Traditional racking methods — 150mm

The following images show traditional racking methods in common use in the UK prior to these tests.

Figure 9.4 – Traditional 150mm tubes laid out on a barge. Note the proximity of mortars to one another

Figure 9.5 – Typical old-style rack of 5 × 150mm mortars

Figure 9.6 – Typical old-style rack of 3 × 150mm mortars

The following images show the destruction of traditional 150mm racks.

Mortar tube square rack steel

The second style of racking tested was a square rack containing nine tubes in a 3 × 3 configuration. The rack consisted of a plywood top with

Figure 9.7 – Firing of 150mm shell in mortar

Figure 9.8 – Damage to tube from which shell was fired

150mm holes drilled through it, allowing the mortar tubes to be dropped through from the top. The base of the frame was made of a steel grid structure that allowed for the separation of the tubes at 200mm distance.

Figure 9.9 – Effect on adjacent tubes and rack

Figure 9.10 – Steel and wood composite mortar racks with separated tubes

Mortar tube square wood rack

The third style of racking tested was similar to the second but consisted of wooden racks top and bottom rather than a steel frame for the base.

Figure 9.11 – Damage observed after firing

Each tube is separated by 200mm in any direction and the rack contains a total of nine tubes. The rationale for this style of rack was to eliminate the steel element and thus avoid the risk of metallic fragmentation.

Figure 9.12 – Wooden rack-style prototype with scaffold pole supports

Figure 9.13 – Rack after firing. Note that adjacent tubes are unaffected by shell burst

Ladder rack construction

The fourth style of racking tested is what we refer to as the ladder rack. The rack consists of two rows of eight tubes separated by 200mm distances between tubes. Two racks side by side total 32 tubes, which is the number of cues of a FireOne module. The rack is relatively lightweight and can be transported flat packed, as the length fits into the rear of a standard transit van.

Figure 9.14 – Typical ladder rack

Ladder rack damage

The following images show the effect on the ladder rack of deliberately bursting a shell within a tube.

Figure 9.15 – Damage to ladder rack after firing. Note that adjacent tubes are unaffected

Figure 9.16 – Damage to ladder rack after firing. Note that adjacent tubes are unaffected

Figure 9.17 – Firing of "end" tube. Note that adjacent tubes are unaffected.

Mortar tube catching methods

Kimbolton Fireworks have also conducted a number of tests to establish what type of catchers can be introduced to further reduce the risk of hazard from a displaced mortar tube.

Figure 9.18 – Herras-type "catchers"

This type of arrangement is only used on certain types of displays, for example on large displays where the finale sequence may consist of large bouquets of shells fired in quick succession from one position.

In this instance a Herras™ fence, surrounding the firing position and oriented according to where the audience is situated, can reduce any hazard. The use of a catcher and the type of construction must be assessed individually by each company.

Recommendations

The overall recommendations and conclusions of this study are:

- Fibreglass mortar tubes are believed to be of sufficient strength to perform the task required for firing mortar shells.
- All maroons should be fired from single mortar tubes, preferably buried and not racked with other shells, especially other maroons.
- Stronger HDPE tubes should be used for multi-break and cylindrical shells and shells with flash or report effects.
- 75mm and 100mm shells do not significantly breach the mortar tube, and during testing did not drastically affect the rack they were contained in.
- 125mm and 150mm shells that are contained in racks should contain separation distances between tubes.
- Open styles of racks similar to the ladder rack, which effectively holds each tube individually, allow the forces involved to dissipate and significantly reduce the risk of adjacent tubes being affected from a mortar failure.
- Traditional-style mortar racks for 125mm and 150mm shells are flawed and will largely result in a failure of the rack, affecting the trajectory of adjacent mortar tubes.
- Mortar tubes wider than 200mm should be individually buried and separated by a reasonable distance.

Conclusions

There is little any of us can do to prevent a shell failure, or indeed the frequency of a shell "detonation" inside the mortar tube. What we are trying to achieve is to minimise the impact of a premature detonation and therefore reduce the risk that such an event might affect adjacent tubes and the racking system.

- The ladder rack is a suitable solution to minimising the risk of a shell failure in a mortar tube causing a destructive failure of the mortar rack and affecting the stability of adjacent tubes.
- The ladder rack design effectively introduces two separate methods of support for the mortar tubes, with the top section and bottom section.
- The tests demonstrated that in the event of a shell failure and destructive breach, only one section is likely to be significantly damaged, not both, i.e., the bottom section may be destroyed in part but the top section will largely remain intact.
- In this instance, even in the unlikely event of major destruction to one section of the ladder rack, the remaining section will provide significant support for the mortar tubes to remain almost vertical.
- The design and construction of the rack has (to the best of our research capability) proven that even with a significant destructive breach the structure of the ladder rack remains intact and the remaining tubes will remain upright and significantly supported.
- The separation distance reduces the risk of one shell failure propagating to an adjacent tube and leading to multiple shell detonations, which could lead to a higher risk of some or all of the tubes in the rack being affected, where their trajectory could compromise the safety of operators or the audience.
- The separation distance introduced also reduces the risk of a live shell contained in a disrupted tube actually being ignited. Even in the unlikely event that an adjacent tube is knocked over, the forces and flames will not breach the tube and ignite the shell contained in the disrupted tube. This is not the case for shell chains linked using pyrotechnic delays or shells independently fired by an automated firing sequencer or computer-fired display.

Chapter 10 – Choosing the right fireworks

Tom Smith – Davas Ltd

The preceding chapters have provided enough information for the designer of the display to have confidence about the types and calibres of fireworks to be used. What remains is the artistic choice of fireworks (which may be limited by supply) and the specific requirements of the client and the event. The following chapter will concentrate on designing the display itself.

Event-specific issues

The event organiser, or the nature of the event itself, may dictate what specific types of fireworks may be required:

- Quiet or loud (dependent on venue and choice)
- Specific colours (to match a theme or product)
- Any local hazards (these should have been considered before this stage)
- Highly specific requirements of the client

As an example of the last, at one display we performed, our proposed use of "serpent" cakes was rejected on account of the client's religious beliefs, which considered such items inappropriate because of the role of the serpent in the story of Adam and Eve.

Noise

Almost all fireworks produce some amount of noise — producing a truly silent firework display is almost an impossibility. There are venues and occasions where the production of noise should be kept to a minimum:

- Displays for young children (although often once the display is started the level of noise can be increased without causing distress)
- Displays where wildlife can be affected (as above)
- During quiet passages of music at pyromusical events

Effects on animals

Animals react to changes in the level of noise, and in particular to sudden explosive events (perhaps reminiscent of gunshot). Observers

Table 10.1 – Noise issues

Firework type	Noise issues
Shells	Lifting charges Bursting charges Shell components (e.g., whistles) Crackle stars
Roman candles	Lifting charges Bursting charges (for bombettes and splitting comets)
Rockets	Motor ascent Bursting charges
Mines	Lifting charges
Fountains	Functioning noise Whistle effects
Set-pieces	Quickmatch between components Additional effects (e.g., whistles or Roman candles)

of displays fired over bodies of water where there are populations of wild birds expect the birds to take flight when the display is started, but are often surprised to find that the birds return to the water part way through the display and are seemingly oblivious to the noise thereafter.

Where possible it is sensible to try and minimise this transition. For example, it has been found that playing music in increasing volume to horses in stables prior to the display is extremely effective in masking the start of the display, thus minimising any possible distress to the animals and, often as importantly, any consequent expense. This approach works particularly well for a "normal" display, which goes on for a period and then stops. It is less effective for the intermittent use of fireworks.

There have been moves by the animal lobby to restrict the noise output from all fireworks to, for instance, 96dB — although exactly where such a measurement should be taken is not explained. The European Standard for consumer fireworks sets a maximum sound level of 120dB (AI max) at the specified safety distance of the particular firework being tested. 96dB is an entirely unrealistic level — it is roughly the same noise level of that of a large book being dropped on a wooden table measured 1m away!

The sound pressure, and hence the perceived sound level, of any firework decays with the square of the distance from the source and hence the farther away from a firework noise source the better. However, local effects, including

- Wind strength and direction
- Humidity
- Topology of the land
- Reflectiveness of local features (buildings, trees etc.)

may well have a more significant impact on the perception of the sound at the receptor than the pure mathematics of sound propagation.

Even quickmatch can produce a loud enough "crack" when lit to upset horses and cattle — in circumstances where such animals are present. But if the display must proceed (and in general it may be preferable to avoid using any fireworks in such situations), there should be a very careful consideration of both the types of fireworks appropriate (usually only ground based), the specific effects (e.g., no crackle) and the fuse used. Slow-burning fuse such as Plastic Igniter Cord (PIC) or Wasag may be quiet enough to be used — but we recommend trialling any fuses and devices before the night!

Site specifics

On some sites the use of particular colours is not appropriate or efficient — this is particularly so when a structure may be bathed in a particular colour of light (often sodium lighting) or when another feature of the event creates a lot of light — for instance a traditional bonfire. In both cases the yellowish light produced renders some of the more subtle gold effects completely useless — the effects tend to look like smoke.

Colours

In some cases the client asks for specific colours to be used — to reflect a theme (e.g., silver wedding) or a corporate image. However, sometimes the desires of the client can be unproductive and a tactful plea to reconsider may be appropriate. We have lost count of the number of times we have been asked to faithfully reproduce a black logo in lancework!

Chapter 11 – Display design

Tom Smith – Davas Ltd

It is probably true to say that the design and construction of fireworks themselves have changed little in the last 50 years — there have been developments in the chemistry of the effects and undoubtedly there will be further, sometimes quite dramatic, changes in this area in the future. However, there has not been development of new fundamental types of fireworks for, probably, hundreds of years. Shells produced today would, for instance, be instantly recognisable as shells to firework manufacturers of a hundred years ago. Similarly, even "cakes" or single-shot devices would be recognised as Roman candles or mines.

The use of fireworks, however, has changed almost beyond recognition from that of our forebears. In general, displays have become shorter in duration, or indeed may be intermittent in nature — interspersed with, perhaps, lasers or lighting, or used to punctuate the event itself. We doubt very much that we will ever return to the display designs of the early 1900s, with elaborate (and expensive) "machines" designed to be seen and admired by the audience before as well as during firing, or the vast lancework designs so admired around the same time.

Other display features however are almost cyclical — the current desire to reduce duration and increase intensity can ultimately become boring, and in time we are sure there will be a reversion to extended display durations with significant periods of low-intensity fireworks (for instance mines, fountains, strobe effects etc.), in part at least to make the finale seem even more spectacular. The variation of intensity of the display is especially important in pyromusical displays — which will be discussed further in Chapter 16.

Fundamentals of display design

Even in a short display it is essential to maintain the audience's attention by providing them with a large variety of effects and the patterns produced by those effects. Where it is possible, for instance when firing from structures, this is relatively easy to achieve; when firing from a restricted site (and especially for small displays) this is much more difficult.

Sequences

The design of many displays can be described in terms of sequences of a particular colour, or a particular effect. For instance, a sequence during a display could build in intensity and height by use of small-calibre

Roman candles, building to larger-calibre candles of the same type, then shells of the same colour fired "over the top". In a noisy sequence the intensity may be increased by using crackling effects, followed by small maroons, followed by larger-calibre and louder maroons.

For smaller displays, especially those that are manually fired, keeping to a consistent pattern of effects can help maintain continuity (i.e., no gaps) and variations in intensity and height without risking boring the audience. A typical sequence of fireworks fired in rotation for a small display could be

- Roman candles or "cake"
- Rockets
- Mines (or set pieces)
- Shells

and ending with a finale in which, for instance, shells and cakes are fired simultaneously.

For large displays, especially where the display is fired electrically, the design can be considerably more sophisticated, and the flow created by the "sequencing" outlined above may be achieved and augmented by precision firing of a variety of effects — both firework (i.e., by different types or of different calibres) and by design (creative use of patterns).

Symmetry vs asymmetry and odd and even numbers

Historically it appears that symmetrical firing around a central firing point has been the norm; however, designers are increasingly choosing to fire from an even number of positions and to deliberately choose to introduce asymmetry into the display design.

Obviously much depends on the firing site layout and the possible use of structural elements into the display. A central feature inevitably dominates the display (for instance see details in Chapter 20 of the London New Year displays fired from the London Eye), whereas displays fired from multiple points may mean that simultaneous firing from each point is most desirable.

A linear (or near linear) arrangement of firing points allows progressions of fireworks to be fired along the points, adding a new dimension to the display that is most readily appreciated by the broadcast audience.

Using very simple diagrams for the firing of mines or Roman candles, it is possible to illustrate a variety of possible firing patterns. These illustrations are not definitive and exclusive, and there is plenty of opportunity for a creative designer to have an enormous variety of patterns and effects during a display.

In the following diagrams (Figure 11.1) we have assumed five equally spaced firing positions centred around the axis of the display (which may be the sight line from the stage to the audience). Firing could take place from any or all of the firing positions.

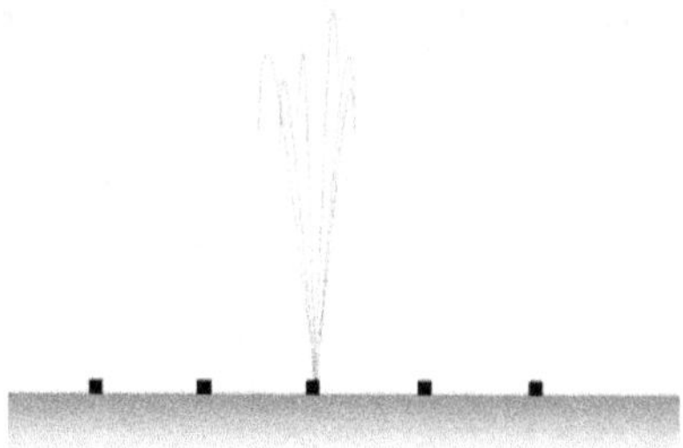

(a) Single firing — vertical

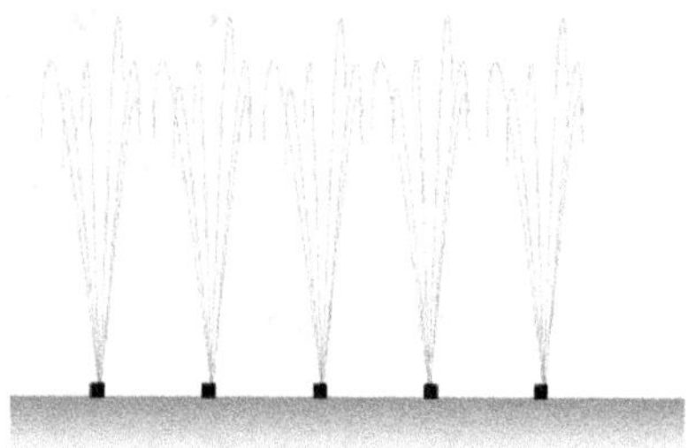

(b) Multiple firing — vertical

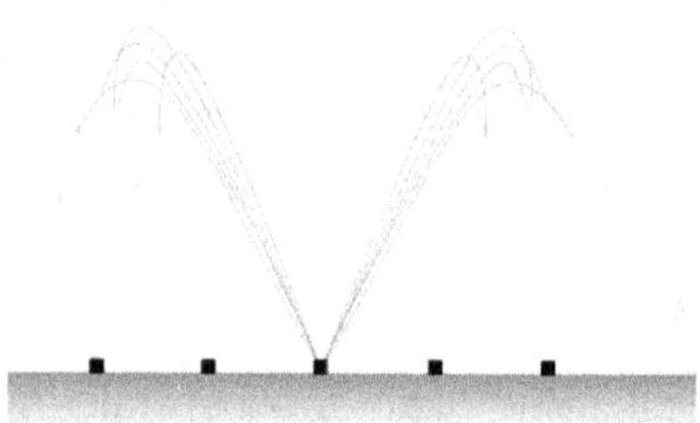

(c) Symmetric firing from central point

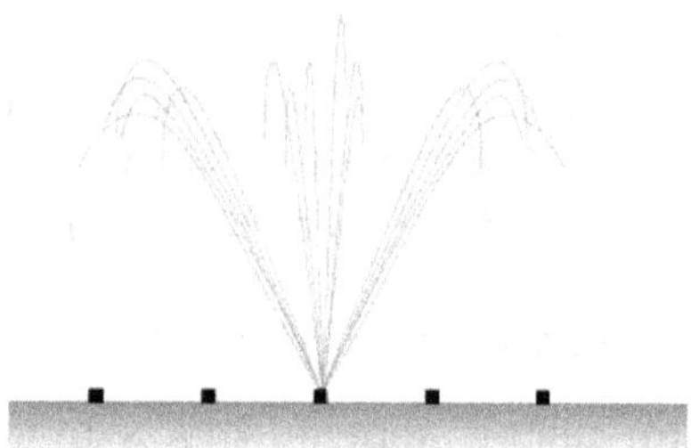

(d) Symmetric firing from central point — for instance 75mm mines

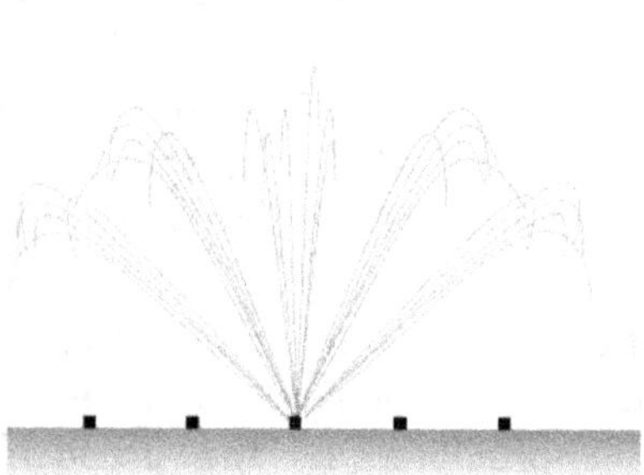

(e) Symmetric firing from a central point — for instance long-exposure photograph of 30mm comet candles (five on a frame)

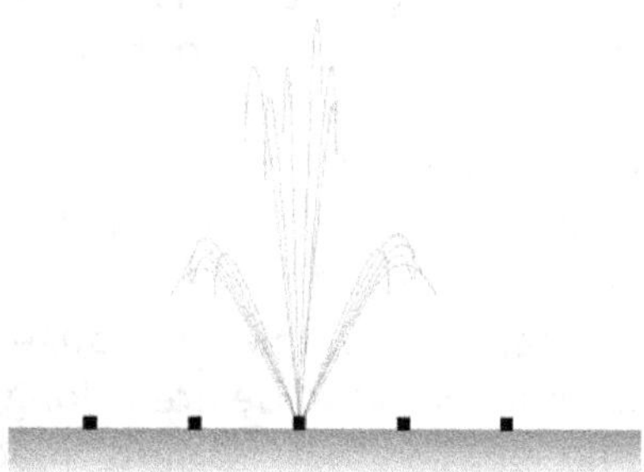

(f) Symmetric firing from central point with different calibre effects — e.g., 2 × 50mm mines and 1 × 75mm mine

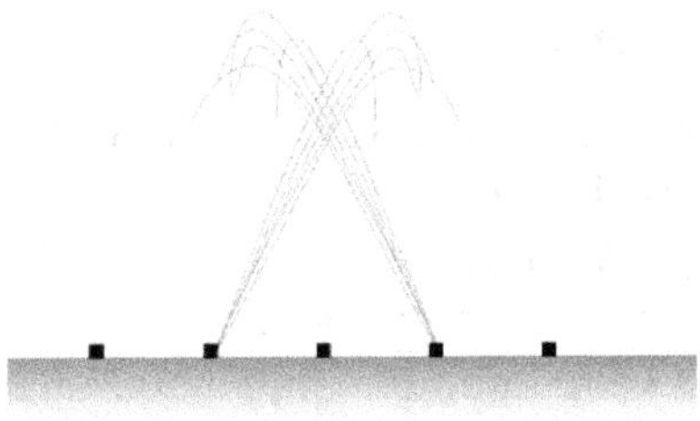

(g) Symmetric firing from two points

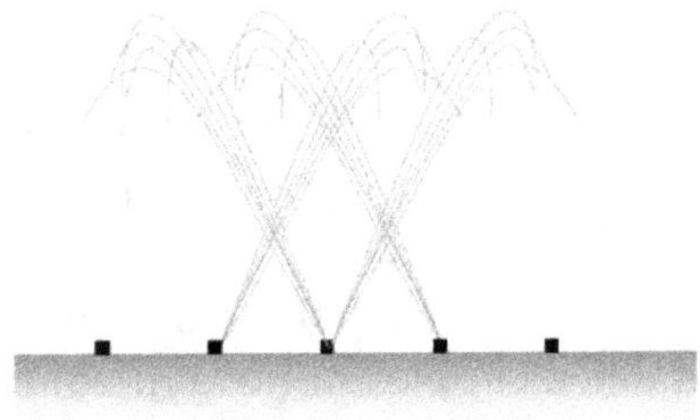

(h) Symmetric firing from three points

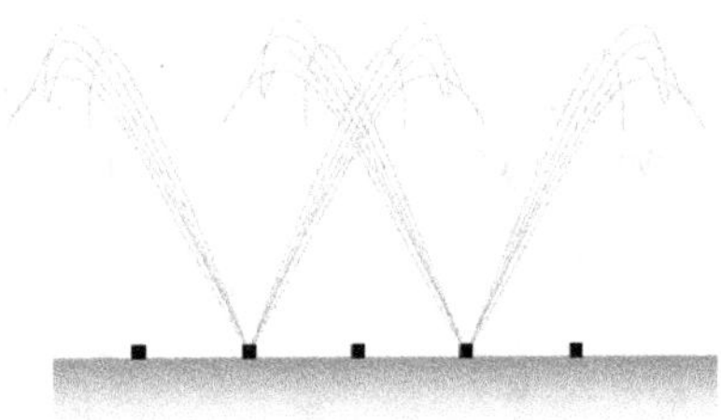

(i) Symmetric firing from two points

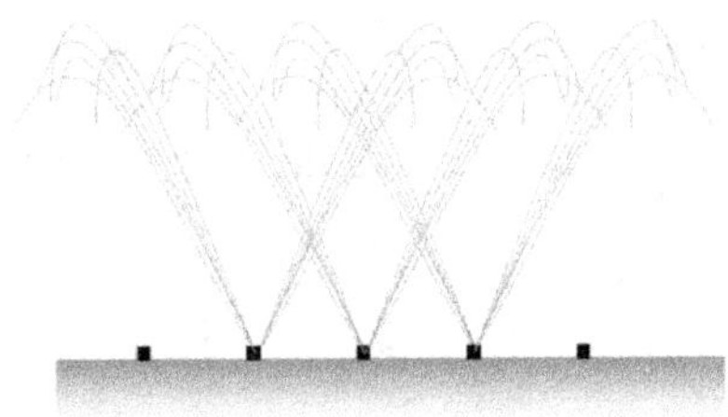

(j) Symmetric firing from three points

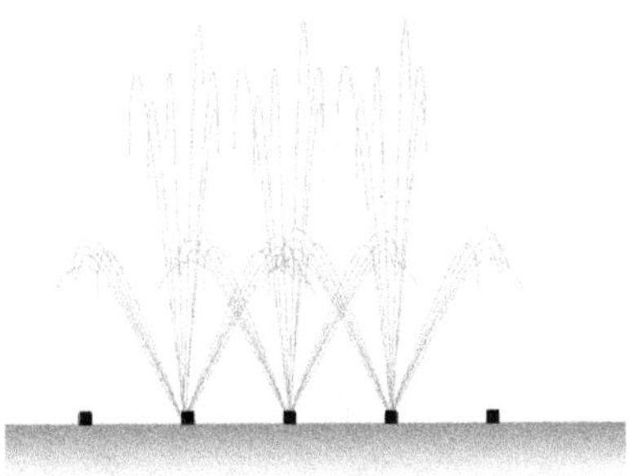

(k) Symmetric firing from three points using different calibre effects

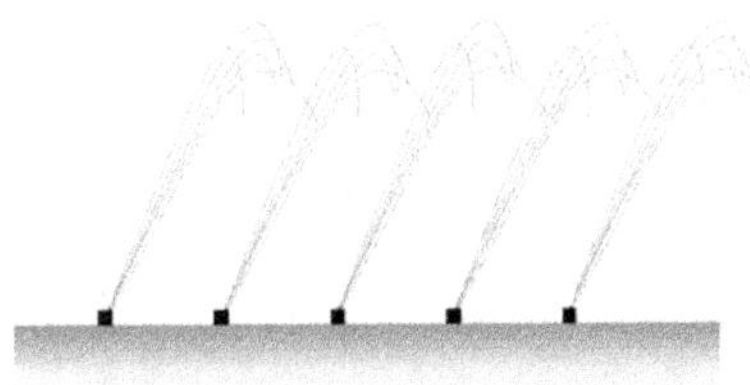

(l) Asymmetric firing from five points — perhaps to provide animation with effect (m) below

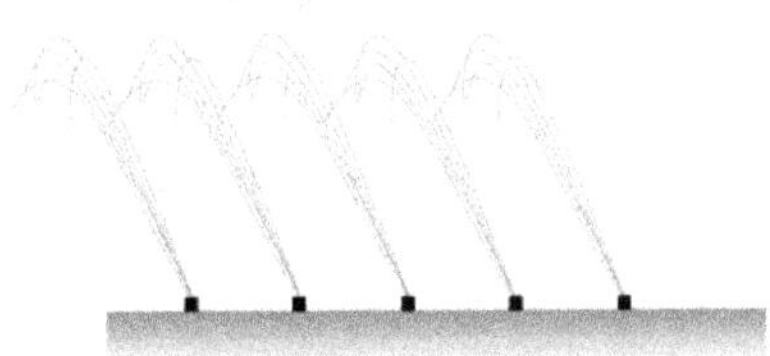

(m) Asymmetric firing from five points — to provide animation from (l) above

Figure 11.1 – Various arrangements of Roman candle or mine effects

Firing fireworks from structures — "chasers"

One of the features of recent displays, particularly when fired from extended frontages or from structures, is the use of "chasers" — a sequence of effects, very often single-shot mines or comets, fired sequentially along the frontage or mirroring the structure. The most notable examples of this type of effect have been on large iconic structures such as the Sydney Harbour Bridge or the Eiffel Tower.

The following are examples of the types of chase effects that can be devised — for clarity they are shown on flat ground rather than on a structure (Figure 11.2).

The chase principle can also be extended to the firing of shells (Figure 11.3), usually fired over a lower-level effect to maintain

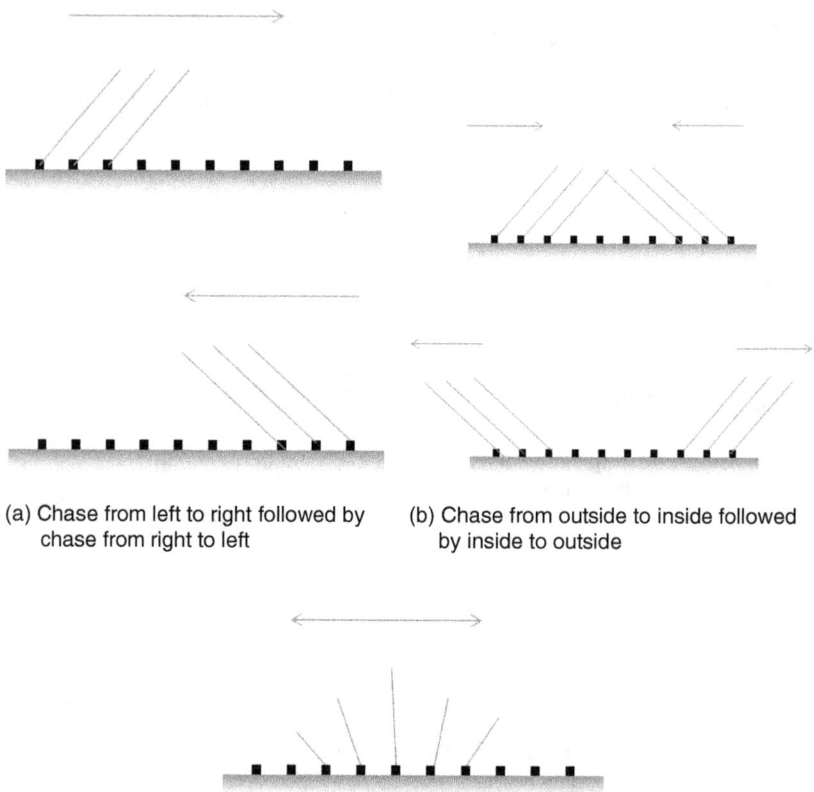

(a) Chase from left to right followed by chase from right to left

(b) Chase from outside to inside followed by inside to outside

(c) Expansion effect from centre to outside

Figure 11.2 – Various "chase" patterns

continuity. This can be very effective in creating variety at relatively low expense — and only single, albeit well-chosen, high-quality shells into the dark space above the low-level effect so that each may be seen to its best advantage. Pattern shells and simple chrysanthemum-type shells are particularly effective in this role.

The more complex and interesting the structure, the more complex and interesting and diverse can be the effects. For example, for a display

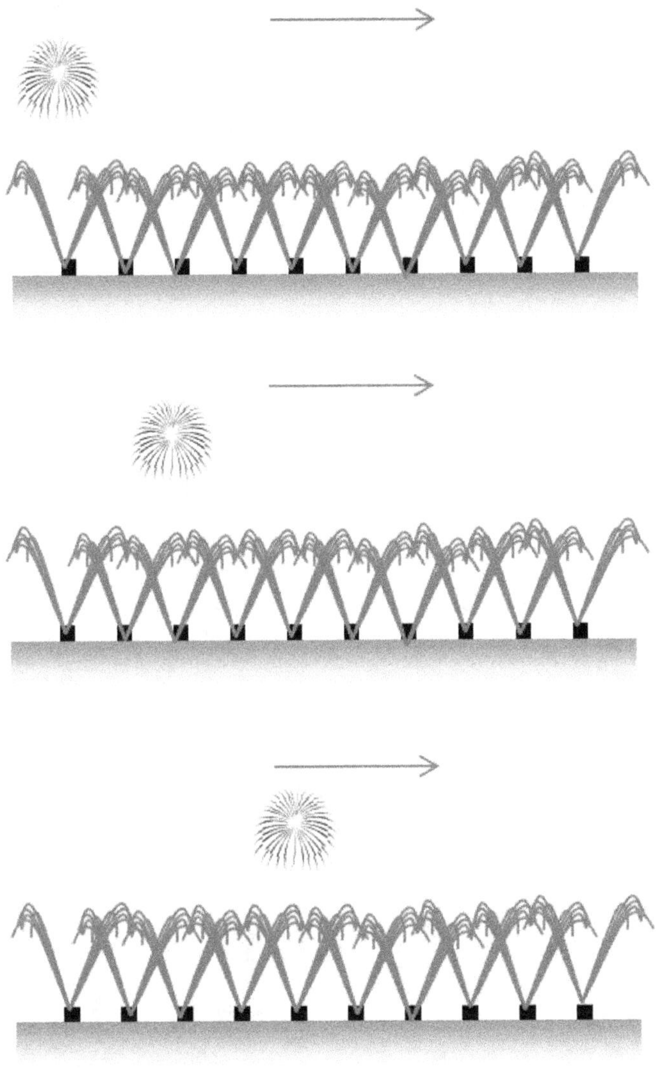

Figure 11.3 – Use of chaser shells

of comets from the London Eye (Figure 11.4) (see also Chapter 20) the following are some of the possible effects:

The timing of firing chasers can vary widely and although there has been a move in recent years to speed up the chases, perhaps more to demonstrate the accuracy of the firing systems rather than pure artistic considerations, the variety in the display design that we strive to achieve

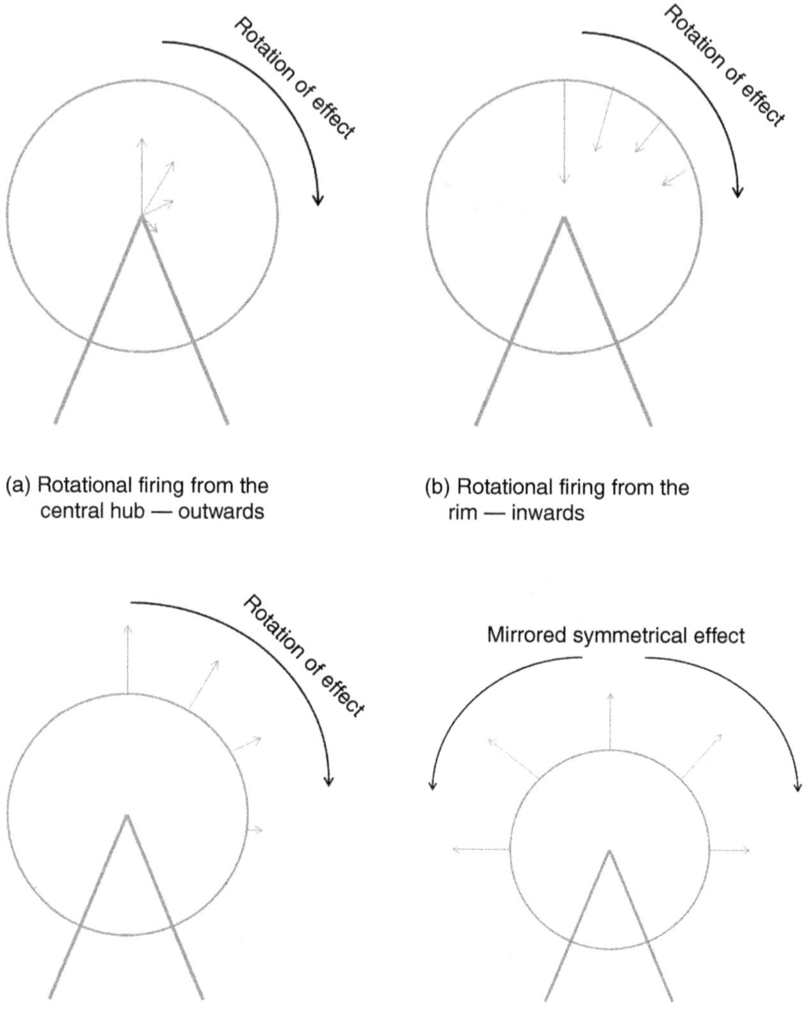

(a) Rotational firing from the central hub — outwards

(b) Rotational firing from the rim — inwards

(c) Rotational firing from the rim — outwards

(d) Mirrored firing from the rim — outwards

Figure 11.4 – Possible effects from a structure — the London Eye

can be made by varying the speed of the chases, from deliberately slow to almost instantaneous.

Still photography is not good for illustrating these "chase" effects, as the timing of the design cannot be seen. Because exposure times are generally quite long (in the range of a few seconds) it is impossible to distinguish, for example, a rotating effect from a single "hit" from all the firing positions simultaneously. However, a number of amateur and professional video clips are available on Internet websites such as YouTube©, and readers are encouraged to seek inspiration from them in planning their display.

Finale planning

The finale of a display is the part of the display that the audience goes away remembering. If the finale of a display is coherent and intense, the audience's perception of the display will be that it was coherent and intense. If, on the other hand, the finale is weak and with many gaps, the audience will go away thinking the whole display was weak and lacked coherency.

The finale is also the one period in the display when concentration is everything — even over-riding, perhaps, the need to avoid too many colours in the air at once — in fact, most people watching will not remember anything else from the last 45 seconds of a display other than the intensity and noise produced!

That is not to say that careful planning and design are not important for this section, just that, perhaps, the priorities are different. In the finale there will almost always be overlap of shell bursts, and precise, expensive effects will stand out less. There will, inevitably, be a large amount of smoke at this point in the display (from what came before or from the finale itself) and it is advisable to avoid bright white effects that tend to light up the smoke and wash out other effects (unless this is a deliberate design feature).

Fireworks in conjunction with other special effects

Increasingly, fireworks are used with other special effects for the production of an event. This may be, for instance, the use of other effects prior to the start of the firework display — perhaps to entertain the audience from their arrival time at the venue until the finale, or in the buildup to the finale of the firework display itself.

For example, at the London New Year displays fired on and around the London Eye, the London Eye itself is lit with coloured lights both within each capsule and externally, and by projection of images on adjacent structures. This is an important part of entertaining the audience on an inevitably cold night from the middle of the evening until midnight. Furthermore, the firework display is designed in conjunction with the lighting designer so that the capsules may be lit with coloured lights in synchrony with the fireworks.

In a similar manner lasers are often used prior to, or in conjunction with, fireworks, although (thankfully in the opinion of the author) the novelty of this is declining. Like so many things, not only in firework displays, when the planning is extensive and coherent and the effect is designed well, the result can be more than the sum of its parts. On the other hand, and all too often, if there is little thought and resulting synchronisation, one effect reduces the impact of the other.

One undesirable consequence of using other light sources, including the traditional bonfires in the UK when lit prior to the fireworks, is that the extraneous light diminishes the visual effect of the subtler colours — especially some of the gold effects made from charcoal — which then appear washed out or simply smoky.

Intermittent use of fireworks

At very large events, for instance at the opening ceremony of the Melbourne Commonwealth Games (M2006) which is discussed later, fireworks are used to punctuate various parts of the overall ceremony. At first thought this use appears to contradict the fundamental desire to keep the display flowing from effect to effect. However, provided the punctuation is deliberate and of sufficient impact to stand alone, the effect can be dramatic. This is particularly easy to achieve when a large number of identical items are fired simultaneously from a large number of positions. In the M2006 display, for example, "hits" of mines and comets were fired from the stadium roof at the end of each segment of the production within the stadium. Further details of this display are given in Chapter 20.

Subtleties

The following sections are personal opinions of the author, and the opinions of other display designers will differ widely. Often the requirements of the client or the site will dictate display design much more than the individual ideas of the designer — but obviously where there is good planning and communication prior to the event there is likely to be a better result, satisfying the demands of all — not least the intended audience!

Variety is the spice of life

The demands of the audience have changed markedly in recent years, and probably the most important changes have been driven by the fact that people see many more, well-designed displays than they ever did. UK displays in the 1960s lasted typically 40–50 minutes, and going farther back the evidence is that they were even longer. During my time as Display Manager at Kimbolton Fireworks the average display dropped from about 20 minutes (in 1989) to 12 minutes (in 1998). I am convinced that if an audience were asked how long a 10-minute display lasted (without looking at their watches) 40 percent would say 5 minutes, and wished it had been more; 40 percent would say 20 minutes, and wished it had been less; and only a handful of people would have correctly guessed the time AND thought the duration was perfect!

The key to keeping people interested is variety:

- Variety of effects
- Variety of colour
- Variety of tempo
- Variety of concentration
- Variety of noise
- Variety of design

Not too much at once

Apart from the finale sequences where "all bets are off", there is a temptation to try and fill the sky with as much as possible for as long as possible. I am not a great fan of the bang, bang, bang, bang, bang, bang, BANG, finish style of display favoured in certain parts of the world — it all becomes too boring and repetitive too soon. Very often "less is more", and very deliberately lowering the tempo, the height, the width or the concentration can make a particular feature stand out.

However, it is often necessary for the designer to emphasise this feature by repeating it — until the audience "gets it" — for instance where a shell sequence "chases" along the firing front, so that the audience does not simply think there has been a misfire.

Shooting into black sky

For some shells in particular firing into a black sky (i.e., where there aren't a lot of other effects happening) can be extremely effective, and this is particularly true for pattern shells where the audience will perceive the pattern differently depending on their position and the orientation of the shell, and for the highest-quality shells where the perfection of the burst and the colour changing of the stars can often be masked by too much else happening in the spectator's relatively narrow viewpoint.

No two members of the audience will see quite the same show — no two will quite see and hear the bursts at the same time. In very extensive audiences those nearer the firing points will not see the different heights that shells rise; instead they will see larger (higher) shells bursting through, rather than above, smaller-calibre (lower) effects.

It is interesting to note that Japanese fireworkers prefer to shoot into the deep blue sky just post-sunset rather than the black of night.

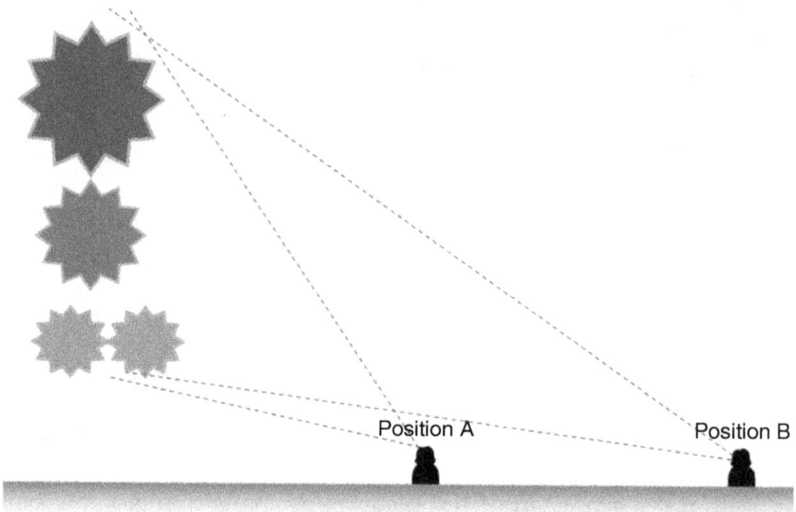

Figure 11.5 – Perspective views of the display (stylised)

Overlaying effects

Again, and particularly for pattern shells, the visual effect and thus the efforts of the display designer and the shell manufacturer can be ruined if there is too much going on in the background. If "normal" shells have to be fired at the same time as the pattern shells, they should be simple (so as not to detract) and of neutral (perhaps gold) or contrasting colour to the pattern shells used so that the pattern is obvious. Firing a sequence of red heart shells against red chrysanthemum shells is never going to be a good idea!

Figure 11.6 – View from the two positions identified in Figure 11.2

Colours and the colour wheel

The familiar colour wheel (Figure 11.7) can be useful to visualise what colours work well together (for instance to produce the maximum contrast) and which do not. However, the light produced by fireworks is usually itself a composite of many colours produced from the emission spectra of a variety of species in the firework flame, and it is often difficult to draw simple conclusions as one would with coloured paints.

It is somewhat meaningless to talk about primary and secondary colours in terms of light emission from fireworks, although there has been some extremely useful work carried out by Beachle[1] and Veline[2] on producing a chemically compatible range of compositions that may be mixed freely to produce a range of intermediate colours. The range of colours is shown on the chromaticity diagram in Figure 11.8, and there are a variety of texts[3] available on the theory and practice of light production from pyrotechnic reactions — but these are outside the scope of this book.

The theory of complementary colours and colour triads is, however, useful in planning a display.

Complementary colours are colours on opposite sides of the colour wheel — for instance red and green, or blue and orange. These mixtures of colours are often found within a single firework — the colours provide the maximum contrast to one another and enhance the visual effect of both. Similarly, firing one complementary colour against another can work well.

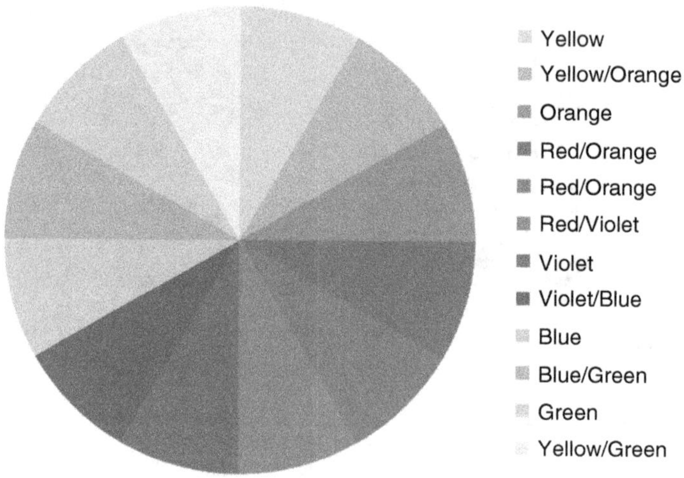

Figure 11.7 – Simplified colour wheel

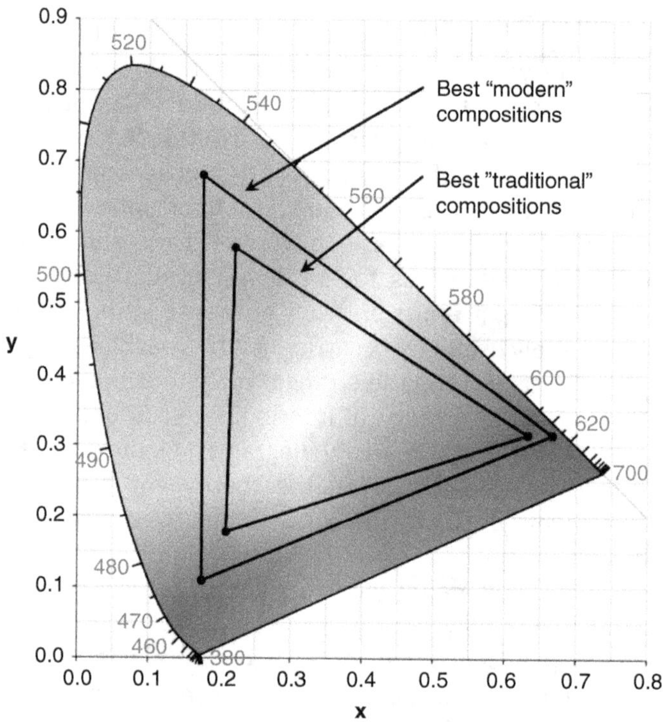

Figure 11.8 – Chromaticity diagram showing the range of firework colours

Triads are three colours equally spaced around the colour wheel — for instance green, orange and purple — and again, such combinations work well within a single firework or between fireworks.

Colours near to each other on the colour wheel generally do not provide sufficient contrast for the viewer to easily distinguish between them, and as a result both can look weak. If colours from the same region are to be used, they should follow one another in sequences and emphasis should be given to ensure the audience appreciates the subtlety of the colour change. In many instances we doubt audiences would actually recognise such a subtle effect!

Number of positions and number of shots from a position

It is not always necessary to fire from all available positions all of the time! The temptation to do so can be great, especially when budgets allow, but the resulting concentration of stars can become boring. Very often changing from firing at every post, to every other post, to every third post and ultimately to only a central post (or the reverse) can add to the variety or concentration and tempo more than any other change, and has the bonus that still images produced at this point are likely to be clearer and more defined than when every available position is used.

As an added benefit, firing fewer items from fewer posts can also mean less smoke, and again this can be a deliberate effect prior to something in the next sequence that the designer really wants to stand out.

Movement along a line and precision placement

Visual variety can be achieved by producing movement in the display, and simple examples of "chaser" type sequences have already been given. It is often tempting to design the "chase" to show off the precision of the firing system (i.e., fire the chase as quickly as possible) rather than to benefit from a more deliberate approach. Slow chases can be used to emphasise a structural feature, or to be a prelude to a series of faster and faster chases (which really show off the skills of the designer and the capability of the firing system). Furthermore, simple slow chases can be an excellent way of preserving budget for the finale! As illustrated above, deliberate slow shell chases, especially with high-quality shells, can also be a way of getting a dramatic effect for minimum expenditure! Each shell should be allowed to burn to completion before the next shell bursts.

Precision placement of shells in the sky is one area of design development that is in its infancy, but some of the most impressive effects we have seen involve careful manipulation of shell-lifting charges and discharge times to produce patterns of shells (not pattern shells).

Finales

Ask the live audience what they particularly like in a show (not just a firework show) and most will comment on the finale — it is the last thing, the bit they go away remembering. This is the one time when the principles of keeping the sky clear, avoiding colour clashes and subtleties of design generally are ignored in favour of high-concentration, high-noise, high-impact fireworks.

Some companies have developed an almost "trademark" approach to finales — often utilising one particular effect (often a long-burning effect like "Kamuro") almost to excess. Although the finale is certainly the one place where excess is good, all too often little effort is put into properly designing the finale sequence and quantity takes precedence over quality of design.

Quantity IS impressive, but quality is more impressive still. Many of the above principles still apply — for instance make the finale maroons move across the sky and finish with a "hit" along the whole width of the display — rather than doing this all the way through the finale sequence.

Avoiding gaps

Except in the largest displays, where intermittent use of fireworks to punctuate parts of the event are designed deliberately, gaps are the worst nightmare of any display designer. In the past there was even a deliberate attempt to create mini-finales and then restart the display when (or if) the audience cheered for more. The demands of the live audience, and their exposure to broadcast or other events, are such that this practice is simply wrong. The danger is (and this is the best case), that even if the audience is clever and the public announcements are clear, no one is sure whether this finale is the "real" finale, or whether there is to be another (and another) if they cheer loud enough. In the worst case they are just left confused and unsatisfied.

Accidental gaps in a display are also undesirable, but they do happen. For hand-fired displays we always recommended holding in reserve a few fireworks that were simple to fire (and simple to stop firing) like rockets, already in a firing frame, and fired by one person dedicated to the job. If they weren't needed to fill a gap, then great — fire them in the finale!

For large shows, especially those fired electrically, the possibility of misfires and gaps is less (although the consequence of an entire line failing are not zero, and the effect can be much more noticeable), so we usually incorporated a few batteries of Roman candles, of neutral colour and preferably of reduced duration, which were on dedicated circuits and ready to fire as needed. In this case if they were not used we did not generally incorporate them in the finales (which were carefully planned) but took them back to fight another day!

Notes

1. J Baechle, *Pyrocolor Harmony: A Designer's Guide*, published by the author, 1989
2. R Veline, *A compatible star formula system for colour mixing*, self published, 1989 — See http://www.skylighter.com/fireworks/how-to/color-fireworks-stars.asp
3. See for example *The Chemistry of Coloured Flame*, K L and B J Kosanke, *Pyrotechnic Chemistry*, Ed. K L and B J Kosanke, Pub: *Journal of Pyrotechnics* ISBN 1-889526-15-0

Chapter 12 – Legal issues

Tom Smith – Davas Ltd

This chapter necessarily focuses on the UK position, but much is relevant to any country adopting the UN Recommendations on the Transport of Dangerous Goods. It is often said that the explosives industry is the second most highly regulated industry after the nuclear industry. This is in part, of course, because of the inherent hazard (not risk) involved with the manufacture, storage, transport and use of highly energetic materials, but it is also in part because of the public perceptions and concerns about the risks involved, and because politicians and bureaucrats desire to manage those concerns and to react to events.

In the UK, and as a result in much of the rest of the world, the basis of explosives legislation was the venerable 1875 Explosives Act.[1] This Act, supposedly drawn up over a single weekend by Colonel Majendie, covered in its original form issues such as storage limits, manufacturing methods, packing and transport and to a limited extent, use.

Finally in 2005, some 23 years after the process was started to revise and modernise the Act, the UK enacted the Manufacture and Storage of Explosives Regulations (MSER), repealed the majority of the Act and replaced it with a regime that was much less prescriptive and placed more emphasis on the hazard's producer doing the managing. Although this approach is more flexible, it is rarely what businesses, and small businesses in particular, actually want or need.

In the same way a variety of other regulations, all made under the Health and Safety at Work Act (the primary UK legislation), have covered packaging, transport, classification and supply. Ironically there is now an initiative to revert to a single set of UK explosive regulations!

In general it is important to consider any accident as a combination of two factors:

- What initiated the accident
- The consequences of that initiation

The initiation event could be external (e.g., a lightning strike), inappropriate (smoking), inherent (bad design or practices) or even unforeseeable, and not all initiations will inevitably lead to severe consequences. Explosives are highly energetic but they usually require direct stimulation to function — it would be perfectly possible (but not sensible) to walk into a fireworks magazine and light a match with no consequences (unlike in a flammable atmosphere).

Once initiation of any explosive has occurred though, there is a great likelihood of spread throughout the entire holding of the explosives within a particular area (e.g., within a store) or across a site. The event will often be rapid and the worst effects relatively localised (unlike the spread of a toxic plume from a chemical factory incident), and in general once started cannot be stopped by external control (e.g., firefighting).

The principles of reducing any initiation and the consequences of any initiation apply equally to the most sophisticated factory environment and the smallest display site.

Summary of UK explosives regulations (specific)

The following is a list of the major, explosives- or fireworks-specific, regulations in force in the UK at the time of writing and which concern display fireworks. In addition to these there are a number of significant general regulations that directly impact on the explosives and fireworks sector, as detailed in the subsequent tables.

Table 12.1 – Summary of UK explosive- and pyrotechnic-specific legislation

Title and date	Acronym	Comments
Control of Explosives Regulations 1991[2]	COER	Limited relevance to fireworks — but requires security measures to be established to prevent losses of any explosives
Manufacture and Storage of Explosives Regulations 2005[3]	MSER	Replaced the 1875 Explosives Acts almost completely. Due for revision in 2011/2. Details the conditions for storage of different hazard types of explosives depending on type of store.
Placing on the Market and Supervision of Transfer of Explosives Regulations 1993[4]	POMSTER	Limited relevance to fireworks — implementation of EU Directive 93/15 on Explosives for Civil Uses.
Fireworks Act 1997[5]		Enabling legislation permitting future regulations on fireworks safety
Pyrotechnic Articles (Safety) Regulations 2010[6]	PA(S)R	Repeals several previous regulations and adopts the EU Directive on Pyrotechnic Articles into UK law
Classification and Labelling of Explosives Regulations 1983[7]	CLER	Adopts the UN classification regime. Largely superseded by transport regulations and due for revision in 2011/12.

Title and date	Acronym	Comments
Explosives Acts 1875[1]	EA75	Small sections of the venerable Explosives Act remain, mainly concerning sales and illegal use — but due to be finally repealed in 2011/12
Identification and Traceability of Explosives Regulations 2010[8]	ITOER	Implements the EU Directive on security of explosives — note these regulations may be delayed until 2013

Table 12.2 – Summary of UK explosives regulations (non-specific to explosives)

Title and date	Acronym	Comments
Control of Major Accident Hazard Regulations 1991[9]	COMAH	Implements the EU Seveso Directives and applies to large sites (e.g., >10te 1.3G fireworks or >200Kg 1.4G fireworks)
Carriage of Dangerous Goods and Use of Transportable Pressure Equipment Regulations 2009[10]	CDGUT-PER	Adopts the European Agreement on the Transport of Dangerous Goods (ADR) into UK law and replaces much UK-specific provisions. However, some derogations and deviations from ADR remain. These regulations replace all the old UK-specific transport regulations.
General product safety regulations[11]	GPSR	General regulations applied to all consumer items
Consumer Protection Act[12]	CPA	General regulations applied to all consumer items
Health & Safety at Work Act[13]	HSWA	Enabling Act of Parliament under which all modern Health & Safety regulations are made. In essence you must not endanger others by your acts or omissions.
Management of Health & Safety Regulations[14]	MHSWR	Applies reporting and structure on companies with over five employees

Of course, there is a wealth of national and international legislation that is not specific to explosives but nonetheless impacts on the manufacture, storage, transport, supply and use of fireworks. For a summary of all UK regulations relevant to pyrotechnics see http://www.eiglaw.co.uk. Readers outside the UK will also find much of relevance although obviously national law varies from country to country.

It may be of great surprise to some readers that there are no specific UK regulations about the **use** of fireworks by professionals. Partly this is for historic reasons; partly it is a function of the over-arching philosophy that persons at work (in essence "professionals") are required to work under a regime of risk assessment and management.

Table 12.3 – Summary of international agreements and regulations

Title and date	Acronym	Comments
ADR[15]		The European Agreement on the transport of dangerous goods (Accord Dangereux Routier). Includes sections on: Driver training Classification Security Vehicle construction Packaging and labelling Placarding Documentation
UN Model regulations and test methods[16]	UN "Orange Book"	The text from which ADR derives. Includes the UN Default Classification regime for fireworks
Seveso directive[17]		This applies to major accident sites only.
Civil uses directive[18]	1993/15/EC	This directive establishes various categories for explosives apart from pyrotechnics and mandates the production of harmonised standards.
Pyrotechnic articles directive[19]	2007/23/EC	This directive establishes various categories for fireworks and other pyrotechnics and mandates the production of harmonised standards

The following sections look at the fundamental principles involved in the various phases in the life of a firework and are intended, where possible, to be entirely country non-specific.

Manufacture

Manufacture of fireworks from raw materials is outside the scope of this book and has been covered extensively elsewhere. The facilities needed for such manufacture are, if it is to be done legally, extensive and complex. The possibility of communication of an explosion from one part of the factory site to any other must be minimised, and in general this is done not only by careful control of the processes involved, but by physical separation of the site areas.

Reworking of pre-made fireworks is a much more common process, often carried out at the display site itself, but increasingly (with good reason but with often over-looked implications) carried out prior to shipment to the display. For instance, the fusing together of individual Roman candles on a frame to produce a bouquet, and the fitting of an electric fuse to the resulting assembly, is often more conveniently done at a licensed facility prior to the display than at the display site where the conditions may not be ideal (weather, pressure of time, lack of materials etc.). However, it is important to consider whether the finished bouquet can then be transported to the display site legally:

- How is it to be packaged? It is unlikely that a package certified for bulk transport of individual Roman candles would also be certified for a collection of the same candles on a wooden or metal frame.
- Has it been classified in the new form and within the new packaging?
- Has the addition of an electric igniter increased the risks of ignition?

Presuming the above issues have been adequately addressed, there are fundamental issues during any act of manufacture, at a factory or on the display site, that must also be addressed:

- Minimising explosive quantities — the lowest amount of material should be exposed at any one time. Completed items should be packaged and removed from the area as soon as possible.
- Keeping explosives not being worked on within their packaging
- Ensuring escape routes are kept clear and unobstructed — for instance, within a factory ensuring that the doors are unlocked and clear, and on a display site never working within the confines of the vehicle
- Prevention of accidental ignition — the most frequent cause of accidental ignition is friction, be it during manipulation (e.g., the cutting of a fuse), or by mishandling. Other potential ignition stimuli include impact (often concurrent with friction — e.g., dropping a tool onto exposed explosive, or dropping a shell with fitted, but unprotected, igniter), electrostatic discharge (either from a person, or from their clothing, or from use of inappropriate tools) and possible water reactivity — mostly during "raw" manufacture — but certain compositions within made-up fireworks could self-ignite on repeated damping/drying out and heating/cooling.
- Ensuring other potential ignition stimuli are removed — it may seem obvious that no smoking should be allowed near fireworks at any time (although the number of occasions the cause of an accident can directly be attributed to it are frightening) but other stimuli should also be removed — for instance electrical sources (sparks), sources of

high-power radio frequency transmission (induction of current in an igniter) etc.

- Keeping the number of people involved to a minimum — not only to reduce the consequences if an ignition occurs, but to prevent the "too many cooks spoil the broth" syndrome.

Storage

The bulk storage of fireworks poses its own problems — although the likelihood of an accidental ignition is reduced (especially once the fireworks are correctly packaged), the consequences of an accidental ignition and spread of fire to the bulk are increased.

Again, the principles apply equally to a factory and a display site, and to long-term as well as short-term situations including at the display site itself:

- Ensure fireworks are kept in closed transport packages when not being used.
- Minimise possible sources of ignition.

Hazards in Storage

Typically, and incorrectly, the classifications determined for transport are used to try to assess and regulate the hazards in storage. In the UK, MSER tries to make an equivalence between what is deemed "Hazard Type" for storage and "Hazard Classification" for transport:

- HT1 = HD 1.1
- HT2 = HD 1.2
- HT3 = HD 1.3
- HT4 = HD 1.4

But the onus is put squarely on the person storing the fireworks (or other explosives) to make a determination of the Hazard Type. The equivalence above is meant to be used as a rough guide. For instance, removing explosives from their (approved) packaging may lead to an increase in the hazard — material that was classified as 1.4G by virtue of strong packaging may present an HT3 hazard (i.e., projectile hazard) once that packaging is removed. In some cases removing from the packaging could conceivably lead to a reduction in hazard.

In extreme cases a transition from HT4 or HT3 to HT1 may occur, especially when the explosives are confined in any way (by the building they are in, the packaging they are in or even by the mass of other boxes surrounding each other). Much discussion has been had about possible "synergistic" effects or specific orientations and arrangements of fireworks in a store that could lead to such an increase in hazard — but there is no scientific evidence, let alone scientific method, to formulate a coherent approach.

Transport

The United Nations Recommendations on the Transport of Dangerous Goods (the Orange Book) is now adopted by the vast majority of countries throughout the world. The text also details the test methods for determining the hazard (not the risk) of a package of explosives. For fireworks the hazard may be assigned to one of five classifications.

Note that whereas all explosives classified as UN0333 present a 1.1 hazard, the corollary is not true — there are many 1.1 explosives that are not fireworks! Hence it is important to detail both the UN number AND the hazard classification.

It is also critical to realise that classification is a measure of HAZARD — no account is taken of how likely it is that any package, or the contents of a package (i.e., the explosives themselves) will ignite or that any single ignition will propagate to the whole consignment. As such any protective features to prevent ignition are effectively ignored by the UN regime, and conversely any features of a device that could actually increase the likelihood of an ignition are accepted *de facto*.

The European Agreement on the Transport of Dangerous Goods by Road (referred to by its French acronym — ADR) is the implementation of the UN Recommendations into the EU, and is the basis for member states' own national provisions. Indeed in the UK the transport regulations effectively just invoke ADR in its entirety, albeit with some relatively minor derogations.

Flash powder issues

One of the most contentious issues in recent years has been the UN's decision to change, and then try to quantify, a definition of flash powder. In part this stemmed from an erroneous perception by enforcers that "the industry" was trying to circumvent the definitions by creating compositions that fell outside them. This is an inevitable consequence of any definition — to draw a line in the sand and then not expect people to come all the way up to that line is unrealistic and naïve.

Table 12.4 – UN classification for fireworks

UN Number and Hazard Classification	Main hazard posed	Comments
UN 0333 1.1G	Mass explosion hazard, potential blast wave, entire contents consumed, to all intents and purposes, simultaneously	A significant number of fireworks are now classified as 1.1G — particularly large shells (>=200mm) and items containing a significant proportion (>25%) of flash powder.
UN 0334 1.2G	Projectile hazard — caused either by the item itself (e.g., a Roman candle) or from the packaging	Larger Roman candles may present a 1.2 hazard.
UN 0335 1.3G	Fiery projections — usually for fireworks intentional stars being projected by the design of the item (e.g., a shell or rocket burst) going beyond the 15m test distance	If an item is designed to throw stars upwards (candles, mines) or radially (shells and rockets) more than 15m in normal functioning, it is likely that it will produce the same effect under test, especially when the structural integrity of the packaging has been compromised.
UN 0336 1.4G	Low-hazard items, whose effects are local	Smaller-calibre fireworks may be classified as 1.4G — although the number "defaulting" to 1.4G is reducing.
UN 0337 1.4S	Hazard largely confined to packaging	No display fireworks are classified by the default as 1.4S — only specialised packagings and full UN tests could demonstrate this hazard.

The UK Health and Safety laboratories in Buxton (HSL)[20] carried out some research[21] on adapting the existing UN Series 2 tests to try and determine if a rapid pressure rise under test would be indicative of a possible mass explosion hazard (i.e., a 1.1 classification) when the explosive substance was within an article (i.e., a firework) and as packaged.

Although academically interesting, this approach is fundamentally flawed — the classification of an explosive "as presented for transport" is, as we have shown, so dependent on the manner in which the articles are packaged that this linking is tenuous at best.

Furthermore, the nature of any substance within the test apparatus is not representative of the nature of the substance in the article — for instance a star may weigh several grams, but the test requires only 0.5g of the substance to be examined "in the form in which it occurs in the firework" — this is contradictory and leads to misleading results.

More fundamentally, the test in itself (at least in its initial form) gave such widely varying results that any conclusions were difficult to draw. It was for this reason that HSL proposed the "cutoff" for flash powder to be set at 4 milliseconds or below under the test conditions, but the UN working group decided, quite arbitrarily, to set the limit at twice that: 8 ms. This is just fundamentally bad science by people unqualified to make such decisions and as a result the arbitrary 8 milliseconds limit has been the cause of much friction and mistrust between the industry and the enforcing authorities ever since.

It is not that a simple (not that it is that simple) test to determine the power of flash powder is a bad idea — it is that the implementation of the test and the extension to classification of articles "as presented for transport" is not well made.

Classification and the UN "default" scheme

Knowing the correct classification for the fireworks is essential. Classification is strictly "the determining of hazard as presented for transport", although it is used (often quite erroneously) as a measure of the hazard in bulk storage or, worse, the hazard in use.

Unfortunately in some parts of the world there is confusion between the classification of an item for transport and its suitability for sale to the public. It may be true that "consumer" fireworks, suitably packaged, can be classified as 1.4G (UN 0336), but that does not mean that anything packaged as 1.4G is suitable for sale to inexperienced consumers.

Thus in general the two are not explicitly related, although it is of course true that items that pose a lower hazard in use will generally pose a lower hazard in transport.

Take the following hypothetical example for 75mm shell maroons (salutes):

- Pack 100 in a strong-walled metallic or fibreboard box — Hazard 1.1G (mass explosion — simultaneous ignition, blast wave etc.).
- Pack 10 in a metal "ammunition" box — Hazard 1.2G (metallic fragments).

- Pack 10 in a fibreboard box — Hazard 1.3G — the first functioning throws a second maroon beyond 15m of the UN test distance where it then functions.
- Pack 1 in a fibreboard box — Hazard 1.4G — effects largely confined to the box, minor debris throw.
- Pack 1 in a VERY large fibreboard box — Hazard 1.4S — effects confined to the box.

In other words, by suitable design and choice of packaging the hazard presented by an explosive may be radically altered — explosive articles, in themselves, do not have an inherent classification!

The UN default classification regime for fireworks has attempted to simplify the process of obtaining a classification for fireworks and to address the vast number of firework types and subtypes available. It is simply not feasible to test every product individually. As a result the UN default regime is necessarily pessimistic in its approach and uses limited evidence to assign a "default" classification to fireworks by generic type and, usually, dimensions of explosive content.

The following table is a summary of the main default classifications[22] for fireworks packed in fibreboard boxes. The reader should consult the current table in the UN Orange Book or in ADR to ensure that they have up-to-date information.

It is important to recognise that deviation from the "default" is possible, but only on the basis of test results. The complexity (and sometimes undesirable consequences) of some packs means that the vast majority of fireworks are classified by the default regime, although the mechanics of obtaining a classification by this route vary significantly from country to country.

Table 12.5 – Abbreviated UN default list for fireworks

Type	Subtype, calibre or other constraint (note 1)	Default HD	Comments
Shells	All report shells	1.1G	
	Colour shells ≥180mm or ≥25% flash	1.1G	
	Colour shells <180mm and with <25% flash	1.3G	
	Colour shells 0mm or ≤60g pyrotechnic composition and with ≤2% flash	1.4G	
	Peanut shell	Worst case applies.	

Type	Subtype, calibre or other constraint (note 1)	Default HD	Comments
Preloaded shells (shell in mortar)	All report shells	1.1G	
	Colour shells ≥180mm	1.1G	
	Colour shells ≥25% flash	1.1G	
	Colour shells >50mm and <180mm	1.2G	
	Colour shells ≤50mm or ≤60g pyrotechnic composition and with ≤2% flash	1.3G	
Shell of shells	>120mm	1.1G	
	≤120mm or ≤25g flash, or ≤33% flash and ≥60% inert material	1.3G	
	>300mm	1.1G	
	>200mm and ≤300mm	1.3G	
	≤200mm containing shells ≤70mm with ≤25% flash	1.3G	
Battery or Combination	The most hazardous firework type determines the classification.		
Roman candle	≥50mm and containing flash	1.1G	
	<50mm and >25% flash	1.1G	
	≥50mm containing no flash	1.2G	
	<50mm and <25% flash	1.3G	
	≤30mm, each unit ≤25g and <5% flash	1.4G	
Shot Tube	≤30mm ID and pyrotechnic unit >25g, or >5% flash and <25% flash	1.3G	This item highlights problems with the present table (there are many others) — there is no default for a shot tube >30mm — (not even 1.1G) and hence such items can only properly be classified by test.
	≤30mm ID, pyrotechnic unit ≤25g and ≤5% flash	1.4G	
Rocket	Flash only effects	1.1G	
	Flash composition >25% of the pyrotechnic composition	1.1G	

Type	Subtype, calibre or other constraint (note 1)	Default HD	Comments
Rocket (cont.)	>20g composition and flash ≤25%	1.3G	This is total weight — including motor, stars and burst charge.
	≤20g pyrotechnic composition, black powder bursting charge and <0.13g flash per report (≤1g in total)	1.4G	
Mine	>25% flash	1.1G	
	≥180mm and ≤25% flash	1.1G	
	<180mm and ≤25% flash	1.3G	
	<150g composition containing ≤5% flash, each unit ≤25g; each report effect <2g, each whistle ≤3g	1.4G	
Fountain	≥1Kg composition	1.3G	
	<1 Kg composition	1.4G	
Sparkler	Perchlorate based >5g per item or >10 items per pack	1.3G	Not used extensively in displays
	Perchlorate based ≤5g per item and ≤10 items per pack OR nitrate based	1.4G	
Bengal stick	Perchlorate based >5g per item or >10 items per pack	1.3G	Not used extensively in displays
	Perchlorate based ≤5g per item and ≤10 items per pack OR nitrate based ≤30g per item	1.4G	
Low hazard fireworks	Throwdowns and snaps ≤1.6mg silver fulminate and other novelty items	1.4G	Not used extensively in displays
Spinner	Composition >20g per item, ≤3% flash and ≤5g whistle	1.3G	Not used extensively in displays
	≤20g composition, ≤3% flash, whistle ≤5g	1.4G	
Wheels	≥1Kg total composition, no report effect, each whistle ≤25g and ≤50g whistles in total	1.3G	
	<1Kg total composition, no reports, each whistle ≤5g and total whistles ≤10g	1.4G	
Aerial wheel	>200g total composition or >60g composition per driver. ≤3% flash, each whistle ≤25g and ≤50g whistles in total	1.3G	

Type	Subtype, calibre or other constraint (note 1)	Default HD	Comments
Aerial wheel (cont.)	≤200g total composition and ≤60g per driver. ≤3% flash, each whistle ≤5g and ≤10g whistles in total	1.4G	
Selection pack	This also applies to other mixed boxes — see below.	The most hazardous firework type determines the classification.	
Firecracker	Each tube ≤140mg flash or ≤1g blackpowder	1.4G	Not used extensively in displays
Banger	>2g flash per item	1.1G	
	≤2g flash and ≤10g per inner packaging	1.3G	
	<1g flash and ≤10g per inner packaging OR ≤1g blackpowder	1.4G	

Note 1: "Flash" = loose flash powder or in report/other effects — see above for discussion.

Mixed boxes

Almost every firework display dispatched to an amateur firer or assembled by a professional company will comprise boxes of mixed fireworks.

Provided sensible constraints are met, then we do not believe it is either unacceptable or illegal to permit the mixed packing of fireworks — indeed we believe that the provisions of ADR (the European agreement on the transport of dangerous goods) explicitly allows fireworks of varying classifications to be packed together. ADR does stress that the resulting mixed box must be classified properly and that, normally, the "worst-case" rule applies — but also states that the onus is on the consignor to ensure that the hazard is correctly identified.

Due to the diverse nature of firework displays it is likely that the fireworks transported to any particular site will be packaged in a unique way — i.e., a mixture of types and calibres within a single box. The alternative of taking a huge number of boxes of single types, removing the requisite number needed, repacking and transporting back the remainder is both impractical and, it can be argued, leads to a large increase in hazard as so much explosive is being transported needlessly.

The UN recommendations and their transposition into ADR are completely clear that mixed packing is acceptable and expected for fireworks, and that in essence the "worst-case" classification applies to the mixed box. However, it is implicit in this approach that the packing certificate for the box into which multiple items are to be packed is flexible enough to allow such mixed packing to take place.

At the two extremes, packaging certificates could:

- Detail explicitly what contents are allowed in a particular box — for instance a particular box could contain 72 × 75mm shells OR 36 × 100mm shells — but not allow for the mixed packing of both types. It is even possible that a particular single packaging mark may be applied to different boxes — but that Box A may only contain 75mm shells while Box B may only contain 100mm shells.
- Allow a high degree of flexibility in packing, perhaps within certain constraints such as total mass, limits on numbers of individual items etc.

The latter approach was developed by the British Pyrotechnists Association (the BPA) to allow a wide range of boxes to be used for mixed packing of items for a display of fireworks or pyrotechnic articles. Other packaging certificates may allow a similar flexibility — but before transporting mixed packages in boxes obtained from suppliers, this should be checked.

However, there are several features that must be considered:

- Is the packaging certificate valid for the mixture of fireworks being transported? This is particularly likely to be of concern if different types are transported together.
- Does the packaging certificate cover items that have been modified (e.g., by affixing to a support)? This can significantly affect the ability of the box to contain the fireworks under test or real-life conditions.
- Is the hazard of the resulting mixed box correctly assessed? In most cases this will be the "worst" of the individual items within a box. For example, if a box contains 28mm and 45mm Roman candles the hazard will most likely be 1.3G — the hazard presented by the 45mm candles. However, an assessment should be made to ensure that this general rule is appropriate.

The use of fireworks

Many of the principles discussed in the previous sections apply to the use of fireworks at the display site, but the fundamentals are repeated here for emphasis.

- Minimise the quantities of fireworks being worked on.
- Minimise the quantities of exposed fireworks.
- Eliminate sources of accidental ignition.
- Minimise the number of people exposed.

Once fireworks are rigged and ready to be fired the risk of communication of fire is greatly reduced — after all, during the normal firing of a display a shell in one mortar when fired normally does not affect the shell in an adjacent mortar. When rigged the most likely incident involving premature ignition is that the firework functions "normally". Of course this can have very undesirable consequences — for instance if a barge containing fireworks is towed beneath bridges!

Electric igniters

One issue of classification, and of hazard and risk, occurs more than any other on Internet forums and discussion groups: Is it permissible or sensible to fit electric igniters to fireworks prior to transport to the display site? The simple answer to this is "No", but as always there is more to it than a simple answer and in many cases it may be demonstrated that the process actually reduces the overall risk of the display.

First, in respect to classification, it must be recalled that classification relates to HAZARD — the effect that occurs if the explosive is initiated. It is not sensible to argue that the presence of a few milligrams of an igniter composition can increase the hazard of any firework. What is increased in all likelihood is the RISK of any initiation occurring. This could be because an igniter is sensitive to initiation by shock, friction or, for example, by radio frequency (RF) induction. Adequate protection of the igniter bead (within a strong-walled tube or within the body of, for instance, a gerb or Roman candle may be appropriate — an igniter bead "loose" in the lifting charge of a shell is not) is essential, and in general the advice from igniter manufacturers is to ensure the igniter leads are "shunted" although opinion on this varies.[23]

Second, the fitting of igniters on the display site in itself introduces risks at that point — risks that may be easier to manage back at the factory (e.g., static, spread of fire to the load).

Overall the advice remains "no" — but there may be situations where there is good reason to do so — and this should be as the result of a proper risk assessment of all the factors, likelihoods and hazards involved.

CHAF trials

Following the Enschede explosion[24] several European authorities instigated work on the hazards posed by fireworks in storage — which have become known as the "CHAF"[25] trials. These trials involved the filling or part filling of standard steel ISO containers with various types of fireworks chosen, one suspects, more for their availability than their relevance.

The results of the trials ranged from the expected (i.e., the behaviour in the simulation of bulk storage reflected the default classification derived for transport) to the unexpected and bizarre.

There are several problems with the CHAF trials, fortunately recognised by those that carried out the trials, but misinterpreted by some of the emergency services and some of the press. The main problem is that the trials were not representative of real life — only in the most exceptional circumstances would an ISO container be used to ship or to store a single type, calibre and effect of firework.

Table 12.6 – Summary of CHAF trials data

Trial	Expected result from default classification	Observed effect	Comments
3776Kg 150mm colour shells	1.1	1.1	
5011Kg × 60g rockets	1.1	1.1	
3765Kg × 60g rockets	1.1	1.1	
5288Kg 75mm colour shells	1.3	1.3	
8661Kg waterfall sticks	1.3	1.1	This is the most unexpected result and is subject to some speculation.
1768Kg waterfall sticks	1.3	1.3/1.4	
4680Kg 75mm bag mines	1.3	1.3	

The "CHAF II" project, proposed but somewhat in suspended animation at the time of writing, attempts to be more realistic in its scope and to use small-scale trials to extrapolate to large-scale storage or transport situations. If such extrapolation is valid, then useful work may come out of these studies — but we are yet to be convinced. The problem is two-fold:

First, the range of products available, their different packing configurations and combinations thereof in bulk storage or in bulk transport is so huge that no test can ever be deemed to be truly representative.

Second, there is no evidence for any of the "synergistic" high-hazard effects that are so feared by regulators. The "worst" result in the CHAF trials was the mass explosion of a full container of waterfalls — totally unpredicted and totally unexplained (although there are suspicions). The danger is that trials are run until the desired "gut feeling" result is obtained and then previous, expected, results are ignored, however many there were!

The current emphasis on fireworks, due in large part to the aftermath of the Enschede explosion, tries to make out that in some way fireworks are different from all other explosives. ALL explosives in bulk potentially behave differently to an extrapolation of their behaviour under the UN testing regime (i.e., classification for transport) where the tests and the results may accurately reflect the inherent hazard in transport boxes but do not relate to the close packing of those boxes in large steel or brick enclosures for bulk transport or storage. It is really no different in other classes of dangerous goods, but behaviour of multiple cartons in bulk is not what the UN recommendations sought to address and hence it is not surprising that it fails to work well in this task. There is no easy answer to this conundrum — but it surely is one that deserves a practical and pragmatic approach.

Supply

The legal position of supply of fireworks varies widely from country to country and it is almost impossible to summarise the worldwide situation in a few paragraphs. In Europe, even with the advent of the new European Standards for Category 1/2/3 (essentially consumer items) and Category 4 (items for "persons with specialist knowledge") it is obvious that national laws will remain and thus determine issues of supply, and ultimately therefore, use. This is not necessarily wrong. Getting a group of experts to agree on common standards for products is hard enough — getting them to agree on supply and use is almost impossible. However, it does present potential future problems of which the Commission is fully aware — they have been warned: the supply in one country (where the control regime is lax) and use in another country (where the supply regime is tighter) will become the norm — in much the same way as it has across state boundaries in the US. Furthermore, as use does not involve supply (if I am putting on a firework display I am not selling a product, I am selling a service) there is an inevitable drive for many smaller, less well-trained and probably less competent operators to import product direct from, especially, the Far East, which is not subject to the testing and product-conformity regimes imposed on product that is "placed on the market".

Notes

1. Explosive Acts 1875 and 1923 — See http://www.eiglaw.co.uk/?p=20
2. Control of Explosives Regulations 1991 — See http://www.eiglaw.co.uk/?p=307
3. Manufacture and Storage of Explosives Regulations 2005 — See http://www.eiglaw.co.uk/?p=309
4. Placing on the Market and Supervision of Transfer of Explosive Regulations 1993 — See http://www.eiglaw.co.uk/?p=304
5. The Firework Act 2003 — See http://www.eiglaw.co.uk/?p=350
6. Pyrotechnic Articles (Safety) Regulations 2010 — See http://www.eiglaw.co.uk/?p=355
7. Classification and Labelling of Explosive Regulations 1983 — See http://www.eiglaw.co.uk/?p=24
8. These regulations are due to be published in 2013.
9. Control of Major Accident Hazard Regulations 1991 — See http://www.eiglaw.co.uk/?p=318
10. The Carriage of Dangerous Goods and Use of Transportable Pressure Equipment Regulations 2009. See http://www.eiglaw.co.uk/?p=347
11. The General Product safety Regulations 2004 — See http://www.eiglaw.co.uk/?p=232
12. The Consumer Protection Act — See http://www.eiglaw.co.uk/?p=342
13. The Health & Safety at Work Act 1974 — See http://www.eiglaw.co.uk/?p=10
14. The Management of Health and Safety at Work Regulations 1999 — See http://www.eiglaw.co.uk/?p=316
15. European Agreement Concerning the Transport of Dangerous Goods by Road (ADR) (Revised every two years) — See http://www.eiglaw.co.uk/?p=251
16. United Nations Recommendations on the Transport of Dangerous Goods (revised every two years) — See http://www.eiglaw.co.uk/?p=238
17. See for summary and links — http://ec.europa.eu/environment/seveso/legislation.htm
18. Council Directive 93/15/EEC of 5 April 1993 on the harmonization of the provisions relating to the placing on the market and supervision of explosives for civil uses — See http://www.eiglaw.co.uk/?p=358

19. Directive 2007/23/EC of the European Parliament and of the Council of 23 May 2007 on the placing on the market of pyrotechnic articles — See http://www.eiglaw.co.uk/?p=360
20. See http://www.hsl.gov.uk/ for contact details
21. D Chapman and K Howard, Journal of Pyrotechnics, 29, 2010, 32–39 http://www.jpyro.com/wp/?p=1176 and links to other texts
22. See http://www.unece.org/trans/danger/publi/unrec/rev16/English/02E_Part2.pdf
23. See http://www.emcconsultinginc.com/docs/ematch1.imm.pdf
24. See for example http://www.aria.developpement-durable.gouv.fr/ressources/17730_enschede_ih_gb_031209.pdf
25. For an introduction see D Chapman, Journal of Pyrotechnics, 22, 2005, 67–70 http://www.jpyro.com/wp/?p=116

Chapter 13 – International standards

Tom Smith - Davas Ltd

The firework industry has always been quite parochial, and the national demands and preferences have dictated the types and effects of fireworks manufactured and used. Much of this is due to public expectation, but in recent years this has changed (with more and more fireworks being seen on international media), helped by the increasing costs of manufacture of bespoke, complex items and the value of items available from the Far East.

In Europe there is also a drive to remove barriers to trade, and hence to standardise products so that those certified in one country may, supposedly, be moved between member states and sold to "consumers" without further, nationally based, certification. This approach does have some merit for true "consumer" items — those intended for the general public, but it is extremely difficult, and potentially very costly, to apply the same logic to "professional" items. This is mainly due to the following factors:

- As noted elsewhere, the diversity of product is very great. For instance, the number of shells from a single supplier with green effects is over 3500 — this makes it essential that a sensible product-grouping regime for testing and certification is adopted.
- The number of any particular item produced may be quite small — so the compliance costs can be quite significant.
- Any attempt to reduce the diversity of items is artistically stifling, with no justification on safety grounds.
- Manufacturers making (or importing from outside the EC) material for their own use and not for supply are not required to have product certified — hence driving down the established manufacturing base in favour of a vast increase in the number of importers.

CEN standards

As outlined in Chapter 3, the European Standards Organisation (CEN) was mandated by the European Commission to produce Standards for a range of pyrotechnic articles following European-wide adoption of the Pyrotechnics Articles Directive (2007/23/EC). The Standards address the following pyrotechnic articles:

- Category 1, 2 and 3 fireworks — those fireworks suitable for sale to the general public and for use, respectively, indoors, in restricted areas (such as gardens) and in more extensive areas

- Category 4 fireworks — for use by "persons with specialist knowledge"
- T1 and T2 theatrical pyrotechnic articles (T1 for "consumers" and T2 for, again, "persons with specialist knowledge")
- P1 and P2 other pyrotechnic articles (including pyrotechnic articles for vehicles)

This book is only really concerned with category 3 and category 4 fireworks. The essential difference between these two categories is as shown in Table 13.1.

In developing the category 4 standards the working groups were particularly mindful that the work:

- Should not restrict the types of fireworks available to "professional" users
- Should recognise the vast diversity of products and should, where possible, aim to group products together for the purposes of testing and certification
- Should not prevent, for instance, stars deliberately falling to the ground — but should provide information to the user to allow them to determine if the product is safe to fire at the particular display site, under the particular conditions of the display
- Should provide product performance information to the user in a consistent and unambiguous manner
- Should recognise that professional users are not the same as "amateur" users and do not need to have the same principles applied to them
- Should recognise that professional firers will use their own equipment for firing
- Should not mandate "safety" distances — the actual distances to be determined by the user, using the provided data and abiding by the local regulations and practices.

Table 13.1 – Category 3 and 4 fireworks comparison

Aspect	Category 3	Category 4
Intended use	By "amateur" firers — people without any mandatory training	"Persons with Specialist Knowledge" — i.e., in common language, "professionals"
Limitations on types	Limited to listed types available during development of the Standard. Does not include, for instance, shells.	Allows flexibility to some extent for new developments

Aspect	Category 3	Category 4
Limitations on size	Restricted by calibre or NEC	None
Labelling and Instructions	Explicit instructions for the user, including warnings and description of effects	Limited "standard" text together with performance criteria that allow the user to determine a "safety" distance — see below
Performance criteria	Dictated by type within the Standard	No explicit criteria or restrictions on performance
Manipulation prior to use		Category 4 fireworks include part made and unfused items that require manipulation on-site prior to use.
"Safety" distance	Explicitly given on the label	Not given — up to the "Person with Specialist Knowledge" to determine using various techniques available
Test methods	Explicit within the Standard	Essentially "fire and measure"

The safety-critical data, which should ideally appear on the label itself, depend on type. For instance, for shells, the data are:

- Burst height
- Burst width (or diameter)
- Dropping height of lit stars (to distinguish, for instance, kamuro shells from colour peony shells)

Other information that may be provided to the user in this case includes mandatory, non-performance, data:

- Net explosive weight
- Sound level (specified at a particular distance)
- Calibre
- Type
- Description of effects
- General warnings
- Details of manufacturer

And other, non-mandatory performance data such as:

- Fuse time
- Flight delay time
- Effect duration

The information is presented on the firework label as a series of parameters (the final form of which will be in the Standards, and readers should ensure that nothing has changed since the publication of this book) in the form "X:Y", where "X" is a letter as shown in the table below and "Y" is the value (with units).

In each case the tests needed to determine the parameters are carried out using a standard set of firing equipment — it is then up to the display company or firer to adapt these "Standardised performance data" to reflect the conditions under which the product will actually be fired. For instance, the burst height of a shell is dependent, to an extent, on the calibre of the mortar and its length and other factors. Provided the "Standard" equipment is known, a company can adapt the actual performance parameters and hence the "safety" distances by the methods outlined earlier in this book (Chapters 5 and 7), based on a knowledge of the Standard data and the effect of other factors as illustrated for shells in Table 13.3 below.

Table 13.2 – Parameters used in CEN Standards for category 4 fireworks and T2 theatrical pyrotechnics

Code	Status	Description	Examples/Comments
A	M	Effect height	
B	M	Sound pressure level	Value @ distance
C	M	Debris distance	
D	M	If effect reaches ground	"✓" if effect reaches ground
E	M/O	Effect range	This parameter is mandatory for aqua fireworks.
F	O	Effect broadness	
G	O	Calibre	If not required in description
H	O	Initial fuse time	
I	Not used to avoid ambiguity		
J	O	Flight time	E.g., time to burst for shells
K	O	Effect time	E.g., duration of stars
L	O	Flash powder content	E.g., for compliance with UN default classification (in weight or in %)
M	O	Gross mass	E.g., for calculating storage limits
N	O	Effect	If not explicit in description or name
N	Not used to avoid ambiguity		
P	O	Classification reference	E.g., CAD reference

Code	Status	Description	Examples/Comments
Q	O	Drift	E.g., for aerial wheels
R	M/O	Overall duration	M for aerial wheels
S	O	If multiple effects	"✓" in case of multiple explosion, followed by the number of explosions
T	O	Radial effect distance	Used mainly for T2 items
U	Not used to avoid ambiguity		
V	O	Burning rate	Used mainly for T2 items
W	M/O	Maximum firing angle	E.g., for fanned cakes
X	Reserved for future use		
Y	Reserved for future use		
Z	Reserved for future use		

*M=Mandatory, O=Optional

Table 13.3 – Possible factors affecting performance of shells

Factor	Effect
Mortar length	The longer the mortar the higher the burst height — although this effect is relatively minor. Very short mortars should not be used in an attempt to reduce the burst height.
Mortar calibre	In general, the better the "fit" of the shell to the mortar, the higher the burst height.
Mortar material	Minor effect — the more elastic the material the lower the shell burst height.
Firing surface	Any recoil of the mortar (for instance into soft ground) will tend to lower the burst height.
Humidity	In highly humid atmospheres the efficiency of the black-powder lifting charge is somewhat reduced.
Quantity and quality of lifting charge	Occasionally the lifting charge is manipulated by the display company to change the burst height — the more the lifting charge (or the better the powder grade) the higher the burst height.
Mortar "dead" volume	If there is significant dead volume (by design or if the shell is not placed at the bottom of the mortar) below the shell, the efficiency of the lifting charge is reduced and the shell bursts at a lower height.

Display companies should be able, therefore, to produce performance adjustment tables for a wide variety of firework types under a variety of conditions and apply them to their own particular methods and firing equipment.

European national standards

Despite the work carried out to harmonise the products and to provide consistent and language-independent labelling outlined above, it is inevitable (and already has been declared to be the case — even before the "harmonised" Standards are complete) that European nations will continue to maintain their own "Standards" for use.

For instance, both the French and German authorities prefer to use a fixed multiplier, based on shell apogee (× 0.8 — so a shell that has an apogee of 100m has an 80m "safety" distance) or shell diameter (× 800 — so a 100mm shell has an 80m "safety" distance), to set "safety" distances for shells fired by professionals. It is very obvious from elementary ballistics or a quick look at Shellcalc© that the discharge of a shell at a wide variety of angles could lead to a shell falling to the ground outside this distance, or bursting so that its stars fall outside this distance at ground level. Of course this assumes some malfunction has occurred. Conversely, of course, if every shell functioned perfectly, was fired vertically upwards, burst at design height, and whose stars were extinguished well above the ground — the "safety" distance would be a few metres!

This is patently nonsense — and it is to be hoped that professional designers of displays will adopt (and modify to suit their particular requirements) the principles outlined in Chapter 5 in such a way that in time this is accepted as "a" (if not "the") way to have consistency throughout the world.

US standards

Firework standards in the US have developed, it seems, in a rather ad hoc manner and with the assumption that maintaining the status quo is more important than having to compromise. This view may seem jaundiced, but it is based on many years trying to persuade US regulators that just because "it has always been done like that" does not mean this is the best way to do it in the future. From an outside perspective some of the US decisions seem bizarre, for instance:

- Safety distances — these are not "safe" although they are often treated almost religiously as such — see Chapter 5.
- The classification of an item should not be related to the ability to sell it — hazard for transport and suitability for sale are only (very) indirectly related if at all — this is the "all dogs are animals but not all

animals are dogs" conundrum. It may well be that, suitably packaged, all consumer fireworks can be classified as 1.4G — but that doesn't make all 1.4G fireworks suitable for consumers.

- Neither should the classification of an item be related to the desire to ship it by air.
- Weight limits should reflect modern units and be rounded to reflect them, not be direct equivalents of obsolete units.

The National Fire Protection Association (NFPA) in the US publishes a number of Standards relating to various aspects of firework manufacture, storage, transport and use.

NFPA 1123[1] "Code for Fireworks Display" is administered, copyrighted and published by the NFPA and is an American consensus standard which, like many NFPA documents, is systematically revised on a three-year cycle.

Other relevant Standards published by NFPA include:

- NFPA 1124[2] – "A Code for the Manufacture, Transportation, Storage, and Retail Sales of Fireworks and Pyrotechnic Articles"
- NFPA 1126[3] – "A Standard for the Use of Pyrotechnics Before a Proximate Audience"

The Standards, despite their title, are not equivalent to either the mandatory Standards in the UK, nor are they regulations. They have no statutory authority unless adopted by the "authority having jurisdiction" (AHJ) on a state-by-state and even region-by-region basis. The Standards, which are widely adopted in the United States however, are deliberately crafted with language suitable for mandatory application to facilitate adoption into law by those empowered to do so.

We do not believe that these Standards permit the flexibility and responsibility of the risk-based approach adopted elsewhere, and in many cases (outside the US) where we have been asked to demonstrate compliance with these Standards in the design and execution of major events it has been agreed by all partners that a risk-based approach is more appropriate.

The major issue we have with ANY Standard that prescribes fixed "safety" distances is that those distances are rarely "safe" at all — they may be a good starting point where the risk is maintained at an acceptably low level — but they are not "safe". In particular circumstances the actual distances can be significantly reduced from these prescribed distances — provided that procedures and equipment have been designed to maintain an acceptable level of risk.

Chinese standards for products

The Chinese authorities have sought cooperation with CEN, the European standard-setting body, and are proposing to introduce quality standards along similar lines, albeit with an emphasis on manufacturing procedures rather than performance of the end product. It is hoped that such collaboration will lead to greater product consistency and for testing and certification to be carried out at the point of manufacture rather than exclusively (and expensively) once the product is shipped.

On the other hand, trying to create and enforce Standards that reflect the diversity of the markets will be extremely difficult unless they are structured in such a way that fundamental principles alone are prescribed — and not the absolute performance and labelling requirements.

ISO standards

If the Chinese work with CEN standards it is inevitable that there will one day be a drive to harmonise firework standards throughout the world. While every industry would make the case that their product is "different" and deserves special treatment, I truly believe that in fireworks it would be impossible to produce standards, especially for "professional" product that would satisfy the needs of the diverse industry with diverse custom and practice. Even at the European level the compromise positions reached by some 26 Member States (even if only about five are truly active participants) end up suiting almost no one and, as indicated above, could even lead to an increase in accidents. The prospect of engaging the rest of the world would lead to a "lowest common denominator" approach that would suit even fewer. At least the inherent bureaucracy in such a process will mean I will be long gone before this "utopia" is realised!

Other firework standards

This list is by no means comprehensive — it merely highlights the Standards used elsewhere that we have found to be useful.

Standard	Comments
Western Australia Outdoor Fireworks – Code of Practice[4]	Covers supply, transport and use. Adopts the UN default classification regime for fireworks. Is prescriptive on "safety" distances and, for instance, mortar lengths.
Netherlands – Code on firework displays[5]	New rules, partly following from the Enschede explosion
Canada – Display Fireworks Manual[6]	Covers transport use, rigging, training etc. and also descriptive sections of basic chemistry, science vs art etc. Is prescriptive on shell "safety" distances and effects of wind etc.

Notes

1. http://www.nfpa.org/aboutthecodes/AboutTheCodes.asp?DocNum=1123
2. http://www.nfpa.org/aboutthecodes/AboutTheCodes.asp?DocNum=1124
3. http://www.nfpa.org/aboutthecodes/AboutTheCodes.asp?DocNum=1126
4. http://www.dmp.wa.gov.au/documents/DGS_CoP_OutdoorFireworksWA.pdf
5. http://wetten.overheid.nl/BWBR0013360/geldigheidsdatum_20-04-2009
6. http://www.nrcan.gc.ca/mms-smm/expl-expl/pdf/fireworks_e.pdf

Chapter 14 – The environmental effects of fireworks

Dr Tom Smith – Davas Ltd

The negative environmental effect of fireworks is an area that has attracted much interest in recent years. This negative interest is probably predicated on a general anti-firework lobby as much as it is on any measurable effect, but of course, that is not to say that such concerns should be dismissed, nor that manufacturers, in particular, should not seek to minimise the potential negative environmental effects from their products. However, such a move should not compromise human safety at any stage of the manufacturing, storage, transport, use and disposal cycle.

Raw materials

The raw materials used in fireworks and other pyrotechnic compositions are diverse. In the most general terms a pyrochemical reaction can be described as

Oxidant + Fuel –> Products + Heat.

And to this generalised reaction may be added a large variety of additional components to produce additional effects. Examples of the constituents of pyrochemical reactions are given in Table 14.1 below. Much more detailed pyrochemistry texts[1] are available.
This book is not intended to be a chemistry manual, and much more extensive works on the chemistry of pyrotechnics are available elsewhere.

However, it is inevitable that some of the raw materials used will pose toxic or environmental concerns (although in general this is an issue related to manufacture, as the raw materials are never released on functioning), but to classify fireworks and other explosives on the basis of these components is to ignore the fact that the explosive effects are probably of the greatest concern, and that those explosive effects are intimately linked to the way the explosives are packaged.

This is the flaw in the REACH[2] and GHS[3] approaches to classification and control of explosives. It does no credit to those who introduce such legislation that they neither understand the subject they are attempting to regulate, nor heed the advice of those who do. Thankfully, most countries implementing REACH and GHS have taken the view that the regulations should not treat explosives as articles containing substances (an analogy would be a bottle of bleach) and should recognise that the hazard posed by explosive articles is intimately and fundamentally linked to the packaging in which they are transported and supplied.

Table 14.1 – Basic pyrochemical components

Component	Used for	Examples	Comments
Oxidant	Providing a non-atmospheric source of oxygen	Potassium nitrate Potassium perchlorate Barium nitrate	
Fuels	To burn in the presence of oxidant to produce heat — either for propulsion (hot gas production), for light emission (e.g., coloured flames) or transmission of fire (fuses)	Sulphur Charcoal Metals (e.g., Aluminium, Titanium) Resins (used as binders and as fuels) Lactose (for cool flames)	
Colour agents	Usually metal salts (not metals) which produce colour by atomic or molecular emission spectra	Cryolite (yellow) Strontium Carbonate (red) Copper Oxide (blue) Barium Nitrate (green) Calcium carbonate (orange)	The colour species in the flame is often MX (M=metal, X=Halogen or Oxygen) — it is not, for instance, copper(II) chloride ($CuCl_2$)
Colour-enhancing agents	Usually a source of chlorine in the flame	PVC Saran	
Metals	To produce sparks in the flame		Metals that react with atmospheric oxygen at high temperatures give better sparks.
Binding agents	To produce stars	Dextrin Resins	
Glitter "delay" agents	To maximise the delay before the glitter "flash" reaction occurs	Barium carbonate	The "glitter" mechanism and related effects are not well understood. There are many texts available[4] for advanced reading.

Component	Used for	Examples	Comments
Whistle components	Produce a high-frequency "whistle" when burned in the presence of an oxidant	Potassium salicylate Potassium benzoate	There are several texts available[5] for advanced reading.
Manufacturing agents	To assist in manufacture — e.g., flow agents		

Combustion chemistry

When fireworks burn, a number of gaseous and solid products are formed. There is a considerable amount of data[6] available on the normal combustion products from burning fireworks in the open, and the following equations are merely indicative of the typical combustion products from fireworks and pyrotechnics.

Blackpowder

There has been extensive research on the burning of blackpowder, and the following simplified chemical equations are not intended to be more than indicative of the type of reactions that occur.

$$4KNO_3 + 7C + S \Rightarrow 3CO_2 + 3\,CO + 2\,N_2 + K_2CO_3 + K_2S$$

$$2KNO_3 + S \Rightarrow K_2SO_4 + 2\,NO$$

$$2KNO_3 + SO_2 \Rightarrow K_2SO_4 + 2\,NO_2$$

$$2NO + O_2 \Rightarrow 2\,NO_2$$

$$2NO_2 + 2S \Rightarrow 2SO_2 + N_2$$

What is important is the realisation that a wide range of solid and gaseous products are produced and that each of these, potentially, has environmental and health effects.

Similar, simplified equations can be written for a range of pyrochemical reactions. Note that these equations are not complete — and, in general, are not balanced — they are indicative only.

Nitrate-based flash

$$16Al + 3Ba(NO_3)_2 + 36H_2O => 3\ Ba(OH)_2 + 16\ Al(OH)_3 + 6NH_3$$

$$10Al + 3Ba(NO_3)_2 => 3BaO + 3N_2 + 5Al_2O_3$$

Coloured compositions (normally organic fuels)

$$KClO_4 + \text{Organic Fuel} => KCl + H_2O + CO_2$$

$$KClO_4 + \text{Metal Salt} => KCl + \text{Metal Carbonate}$$

Metal sparks (M = general metal)

$$KClO_4 + 4M => KCl + 4MO$$

$$2M + O_2 => MO \text{ (sparks in air)}$$

Smokes (and other organic fuels)

$$C_{12}H_{22}O_{11}.H_2O + 3KClO_4 => 3KCl + 11H_2O + 4CO + 5C + 3CO_2 + H_2O$$

Metal salts (e.g., Antimony sulphide) (M = general metal)

$$2KClO_4 + M_2S_3 => KCl + K_2SO_4 + 2MO$$

Heavy metals

In the proceedings of the International Pyrotechnics Symposium, the Disney Corporation published a paper[7] on investigating, over a period of many years, the deposition of heavy metal salts into Lake Buena Vista from the nightly firework displays held there for more than 25 years.

In essence the conclusions were as follows:

- There has been very significant deposition of heavy metal salts in the lake.
- The metal salts tend to sink to the bottom and are adsorbed onto the muddy substrate.
- There has been almost no variation of identified flora or fauna in the lake from when the research began.

A recent study in Austria[8] describes "exorbitant" concentrations of barium and other metal salts in the snow following a New Year's Eve celebration. However, the base levels of heavy metals in snow are, unsurprisingly, rather low and hence the conclusions must be treated with a degree of caution. No information was provided on any effects post-melt.

Perchlorates

The use of metal perchlorates (usually potassium perchlorate) as a substitute for chlorates is quite widespread and for good safety reasons. Compositions of perchlorate and metals are significantly less sensitive to many stimuli (friction, impact, spark) than their chlorate equivalents. A number of papers[9] have been published highlighting the concerns about perchlorates leaching into groundwater, and the subsequent toxicological effects on wildlife, including humans. In the main this is a concern only for manufacturers of pyrotechnic compositions where bulk perchlorates may be handled and precautions must be taken to prevent groundwater contamination. Users of fireworks should not be threatened with the same potential controls for a variety of reasons:

- The quantities are low.
- Fireworks are designed not to release their compositions — instead it is the combustion by-products that are released on functioning.
- Residual perchlorate is not a significant by-product of combustion — instead the oxidant essentially decomposes fully on functioning.
- Natural attenuation[10] is an important factor in lowering any perchlorate contamination.

Of course precautions need to be taken if any of the above are not in place — for instance if a firework is damaged. However, the overall risk is extremely low.

REACH

In the European Union the REACH provisions (Registration, Evaluation, Authorisation & Restriction of Chemicals) identify hazardous chemicals and ensure downstream users are identified. The regime is not intended to affect end-users of products.

As in the case of perchlorates outlined above, the provisions of REACH fundamentally misrepresent the mode of action of any explosive. A pyrotechnic rocket is not a carrier for the chemicals it contains in the same way, for instance, that a bottle of bleach is a container for an aqueous solution of sodium hypochlorite.

All pyrotechnics and explosives function by converting the chemicals they contain into combustion by-products, and in so doing generate gas, heat, light, sound or a combination of these, which ultimately lead to the observed effect.

Conclusions

While it must be the case that environmental effects should be considered at all stages of the planning of a display, including the manufacturing processes, it is clear that there is no compelling evidence of the effects of fireworks causing environmental or human health problems if used properly. There is considerable worldwide evidence that the misuse of fireworks (deliberate or accidental) causes many injuries and a number of fatalities each year, and although there is always a predictable "knee-jerk" reaction to any such accident the solution is not to ban fireworks!

In Chapter 21 we discuss some of the new chemistry that, in time, may lead to a lower potential environmental impact of fireworks. However, the balance between human safety and environmental impact and human enjoyment must never be dismissed. Oscar Wilde, who obviously enjoyed fireworks,[11] would certainly have had something to say on the matter!

Notes

1. See for example Chemistry of Pyrotechnics 2nd edition, J A Conklin & C J Mocella Pub: CRC Press (2010), ISBN: 978-1-57444-740-8
2. Guidance on REACH is available from http://www.eiglaw.co.uk/?p=644
3. The Globally Harmonised System for Classification and Labelling of Chemicals — See http://www.eiglaw.co.uk/?p=671
4. C Jennings White, Journal of Pyrotechnics, 8, 1998, 53–70 — See http://www.jpyro.com/wp/?p=291
5. W R Maxwell, Journal of Pyrotechnics, 4, 1996, 37–46 — See http://www.jpyro.com/wp/?p=310
6. See for example I von Maltitz in Pyrotechnic Chemistry, Ed. K L and B J Kosanke, Pub: Journal of Pyrotechnics ISBN 1-889526-15-0
7. T A DeBusk, "Environmental effects of fireworks on bodies of water". See http://www.isfireworks.com/Proceedings/Montreal%20proceedings.pdf
8. G Steinhauser, J H Sterbaa, M Fostera, F Grassa and M Bichlera, Atmospheric Environment, Volume 42, Issue 37, December 2008, pp 8616-8622. doi:10.1016/j.atmosenv.2008.08.023
9. "The Fallout from Fireworks: Perchlorate in Total Deposition", Jennie Munster, Gilbert N Hanson, W. Andrew Jackson and Srinath Rajagopalan, Water, Air, & Soil Pollution, 198, 149–153, DOI: 10.1007/s11270-008-9833-6. See http://www.springerlink.com/content/xt68w81615017458/

10. "Perchlorate Behavior in a Municipal Lake Following Fireworks Displays", R T Wilkin, D D Fine and N G Burnett, Environ. Sci. Technol., 2007, 41 (11), pp 3966–3971. DOI: 10.1021/es0700698 http://pubs.acs.org/doi/abs/10.1021/es0700698
11. The Remarkable Rocket, Oscar Wilde (1854–1900). See also http://classiclit.about.com/library/bl-etexts/owilde/bl-owilde-rock.htm

Chapter 15 – Firework competitions

Tom Smith – Davas Ltd

How to approach a competition

Firework competitions have been established for many years and provide not only an opportunity for the firework manufacturer or display designer to show off their skills, but afford the audience the opportunity to see the art of firework displays at its best.

The general principles of display design apply, of course, to a competition, and the design of the resulting display should seek to engage the audience without repetition (but with sufficient variety) to leave the audience wishing there had been more!

Of course the display design also depends to a large extent on the constraints put on the display designer by additional factors, including:

- The nature of the competition (with or without music for instance)
- The rules of the competition, including possible constraints as to types and calibres permitted or excluded, total cost or total NEC/gross mass
- The duration of the display and any judging penalty for over-running or under-running
- The nature of the judges (technical or lay-persons) and the judging criteria
- Any other constraints (e.g., the requirement to include particular effects or particular colours)

Planning a display for a firework competition is little different from planning any other display, although there is often a great deal more thought and care taken (especially by smaller companies) than for their "run of the mill" displays of a similar scale.

It is vital not to forget that fireworks are a subjective art, and no one can compete assured that they will triumph! Much depends on how the judging is organised, and the balance of firework experts and lay-persons doing the judging.

Plymouth – the British Firework Championships

Since the mid-1990s The Event Services Association (TESA)[1] has organised the British Firework Championships in Plymouth, an event that has grown significantly in recent years and now attracts some 250,000 people spread over two nights of displays.

Each night three competitors each fire a ten-minute display from the Mountbatten Breakwater, an old flying boat terminal superbly situated in the middle of Plymouth Harbour in the southwest of the UK. The breakwater is a concrete stone structure, some 300m long and about 10m wide.

Figure 15.1 – Mortar racking at Plymouth

Each competitor is given a fixed sum of money to spend on the display and transport, and in addition the competitors are provided with local storage and accommodation. In order to ensure the displays are fair, each competitor must supply to the adjudicator of the competition a list of fireworks to be used and their total weight and net explosive content. In recent years the maximum allowable NEC has been fixed at 500Kg; compliance with the declared list of fireworks is checked on-site. The adjudicator also checks the list of declared fireworks against standard data to confirm the NEC and also checks the fireworks against the List of Classified Explosives and Fireworks (LOCEF) to ensure the fireworks are legal in the UK.

The last significant constraint put on the competitors is timing — over-running or under-running the display is penalised on a sliding scale.

Judging

Six judges (three from within the events industry and three from local media) are briefed by the Adjudicator and the judging is broken down into a variety of sections to facilitate comparison between the different displays. In essence these are:

- Range of products
- Rhythm of firing
- Originality of design
- Overall effect

It is important to appreciate that there is no "correct" mark! One judge may particularly like a display with highs and lows, while another may prefer a more continuously intense display. One may prefer emphasised colour changes, while another may prefer a variety of colours overlaid on each other throughout the display. Provided that each judge is consistent, the overall result can truly be taken as an appreciation of the widest appeal of the displays.

Technical marks are awarded by the adjudicator and the organising committee to include:

- Compliance with the competition rules
- Timely and accurate provision of information
- Site layout and safety issues
- Any timing penalties

An audience appreciation factor is also included, although this has proved difficult even with modern SMS technology — mostly it is felt due to the feeling that the last show is always the best!

One plea we always make as adjudicators of the event is to mark the first show realistically. There is a natural tendency of people to judge the last show they have seen better than the show seen previously, and if the first show of the six is marked nine out of ten there is little room for improvement! Over the years the judging process has also been refined to weight the judges' scores to ensure no vote rigging is happening, and to fairly reflect the technical marks (for rules compliance, timing etc.), which are awarded by the adjudicator and the organising committee objectively.

The display is not fired to music, partly because of the practical issues of getting synchronised sound to the enormous and disparate viewing areas, and partly because this competition is realistically billed as a "Firework

Competition". The organisers have decided to let others stage pyromusical events, and to keep the Plymouth competition tightly focussed on allowing each competitor to show off their firework design skills to their best.

Results

The winners of the competition — including the "Champion of Champions" event in 2006 comprising the previous eight champions and celebrating ten years of the event — are shown below:

- 1997 – Fantastic Fireworks
- 1998 – Northern Lights Fireworks
- 1999 – Happy Dragon Fireworks
- 2000 – Pyro 1
- 2001 – Shellscape Pyrotechnics
- 2002 – Shellscape Pyrotechnics
- 2003 – Jubilee Fireworks
- 2004 – Alan Hillary Pyrotechnics
- 2005 – MLE Pyrotechnics
- 2006 – "Champion of Champions" — Jubilee Fireworks
- 2007 – Selstar Fireworks
- 2008 – Pendragon Fireworks
- 2009 – Phoenix Fireworks
- 2010 – Star Fireworks

Further information about the Plymouth site and the displays is given in Chapter 19.

Figure 15.2 – Jubilee Fireworks at Plymouth.
Photo: Jubilee Fireworks

Montreal

The Montreal International Fireworks Competition is probably the most prestigious and long-running competition of its type in the world. Many other venues have adopted the format of the displays, although few can afford to provide competitors with the semi-permanent facilities and extended time schedules as Montreal.

Figure 15.3 – Montreal permanent mortar installation with young Darryl Fleming. 1993

In 1993 and 1996 when the author attended, the display itself was fired on and around the lake at the "La Ronde" amusement park on an island in the St Lawrence River, and comprised three main areas:

- Permanently installed mortars within sand-filled troughs — five positions (and intermediates) suitable for firing of up to 12" shells
- Sandy area adjacent to the lake, suitable for firing small-calibre shells (75 and 100mm) and all Roman candles, mines etc.

- Floating pontoons, suitable for Roman candles, mines and set-pieces as well as being a launching platform for aquatic fireworks

Each of these areas was connected by permanent multi-core wiring systems terminating in a purpose-built, glass-walled firing box on the audience side of the lake — so that the audience could see the show being fired!

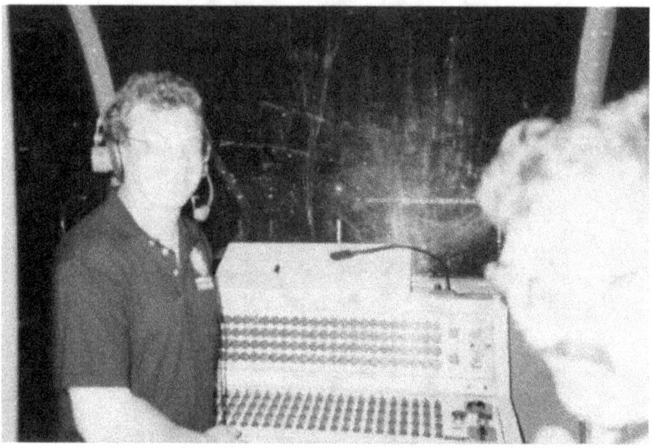

Figure 15.4 – The firing panel in Montreal 1993 with a very young Tom Smith and pyrotechnician extraordinaire Ray Labrie. Note the comparative simplicity of the manual firing system!

A formidable team of Canadian and other firework experts assisted each display company in preparing and rigging their display, and coordinated the rigging and firing of a series of shows throughout the summer.

The displays

The Montreal competition is set to music, which is selected by the competitor. Over the years there have been several approaches to music selection, but like fireworks themselves the appeal of music is incredibly subjective.

What is less subjective is the precise synchronisation of the fireworks and the music, and companies will manufacture, or have manufactured, items with specific burn times to exactly fit the relevant music segment. Some competitors also chose to deliberately break down the overall display into a series of "tableaux", each with a distinct theme (though linked to an overall theme perhaps), and introduced a deliberate pause between each to allow the viewer to "catch their breath" and more fully appreciate the variety and design — obviously others chose to flow one segment into another.

Judging

Montreal establishes a panel of judges, and a proportion of these judges will attend every event. It is incredibly difficult to try and be objective about fireworks, but Montreal has managed to achieve the necessary objectivity by ensuring the pool of judges is large enough that some continuity from show to show is maintained, while at each display bringing in "new blood" to counteract the instinct to judge the latest show better than the one seen a week before!

Figure 15.5 – Montreal lakeside 1993

Other competitions

There are now a large number of firework competitions (with or without music) staged all around the world, and the demands to produce an original and a spectacular display are becoming increasingly difficult. Perhaps it is time to simplify and to allow firework designers an opportunity to demonstrate their creativity, rather than to be almost forced to "stuff" the display with the largest possible quantity of fireworks in order to stand even a chance of winning! Competitors rarely make any profit from a firework competition.

Table 15.1 – Table of Montreal winners[2]

Year	Gold jupiter	Silver jupiter	Bronze jupiter	Sound track and special jupiter
1985	*Pyromusical:* Société Étienne Lacroix (France) *Classic:* Marutamaya Ogatsu (Japan)	*Pyromusical:* Industria Panzera (Italy) *Classic:* Zambelli Int. (USA)	Pirotécnia Igual (Spain)	
1986	*Pyromusical:* Pirotécnia Caballer (Spain) *Classic:* Dongguan Fireworks (China)	*Pyromusical:* Éphémère (France) *Classic:* Garden State Fireworks (USA)		
1987	Austin Fireworks (USA)	Lünig Feuerwerk (Germany)	Pirotécnia Igual (Spain)	Lünig Feuerwerk (Germany)
1988	Pirotécnia Igual (Spain)	Lünig Feuerwerk (Germany)	Austin Fireworks (USA)	Austin Fireworks (USA)
1989	Lünig Feuerwerk (Germany)	Pyrodyne Aerial Productions (USA)	Ex aequo: Pirotécnia Caballer (Spain) and Ampleman (Canada)	Pyrodyne Aerial Productions (USA)
1990	Société Étienne Lacroix (France)	Pirotécnia Caballer (Spain)	Les Artistes Associés (Switzerland)	
1991	Pyrotechnology Inc. (USA)	JNS Pyrotechniek (Holland)	Pirotécnia Zaragozana (Spain)	Pyrotechnology Inc. (USA)
1992	Sunny International (China)	Rozzi's Famous Fireworks (USA)	Pirotécnia Igual (Spain)	Sunny International (China)
1993	Pirotécnia Caballer (Spain)	Sunny International (China)	Ruggieri (France)	Kimbolton Fireworks (UK)

Year	Gold jupiter	Silver jupiter	Bronze jupiter	Sound track and special jupiter
1994	Performance Pyrotechnic Associates (USA)	Syd Howard Fireworks International (Australia)	Marutamaya (Japan)	Imposed theme: Performance Pyrotechnic Associates (USA) Giovanni Panzera
1995	JNS Pyrotechniek (Holland)	Société Étienne Lacroix (France)	Pirotecnia Soldi (Italy)	Jean Émond (SPÎ)
1996	Performance Pyrotechnic Associates (USA)	Pirotécnia Igual (Spain)	WECO Pyrotechnische Fabrik (Germany)	Pierre Walder
1997	Ipon S.R.L. (Italy)	WECO Pyrotechnische Fabrik (Germany)	Günter Vogler (Austria)	Concept Fiatlux (Canada)
1998	Performance Pyrotechnic Associates (USA)	Ipon S.R.L. (Italy)	Pirotécnia Ricardo Caballer (Spain)	Richard Shaw
1999	Performance Pyrotechnic Associates (USA)	Ampleman Pyrotechnie. (Canada)	Pirotécnia Igual (Spain)	
2000	WECO Pyrotechnische Fabrik (Germany)	Pirotécnia Caballer (Spain)	Syd Howard Fireworks International (Australia)	Rothman Benson and Hedges
2001	Pirotécnia Igual (Spain)	Foti's International Fireworks (Australia)	San Tai Fireworks Ltd. (Taiwan)	
2002	Société Lacroix- Ruggieri (France)	Luso Pirotecnia (Portugal)	Ampleman Pyrotechnie. (Canada)	
2003	Royal Pyrotechnie. (Canada)	Explosive Entertainment International (Australia)	Pyromagic Productions Ltd. (Hong Kong / China)	

Year	Gold jupiter	Silver jupiter	Bronze jupiter	Sound track and special jupiter
2004	Platinum Jupiter WECO Pyrotechnische Fabrik (Germany)	Honourable mention Sunny International (China)	Honourable mention Société Lacroix – Ruggieri (France)	
2005	Fuegos Artificiales Júpiter (Argentina)	Rozzi's Famous Fireworks (USA)	BEM Feux d'Artifice (Canada)	
2006	Melrose Pyrotechnics (USA)	Brezac Artifices (France)	Foti's International Fireworks (Australia)	
2007	Pains Fireworks (UK)	IP (Germany)	Pyrospectaculars by Souza (USA)	
2008	Pyrotecnico (USA)	Howard and Sons (Australia)	Sunny International (China)	
2009	Royal Pyrotechnie. (Quebec / Canada)	Pyromagic Productions (Hong Kong/China)	Meltrose Pyrotechnics (USA)	Royal Pyrotechnie. (Quebec / Canada) Green initiative: Foti's International Fireworks (Australia)
2010	Fireworks Spectaculars Canada (Alberta / Canada)	Göteborgs Fyrverkeri Fabrik (Sweden)	Brezac Artifices (France)	Göteborgs Fyrverkeri Fabrik (Sweden) San Tai Fireworks (Taiwan) Public Choice: Fireworks Spectaculars

In fact in most cases they will make a significant loss. While this may be good for the venue (in that they are getting more than they really paid for) it is not a sustainable long-term position to take and ultimately will not benefit the industry or the commissioners of displays.

We would urge competition organisers to try and ensure fair play by adopting some of the techniques we have instituted in Plymouth and elsewhere, and in particular:

- Limit the quantity (NEC) of fireworks to be used.
- Put strict time limits on the displays.
- Have a judging system that is as objective as possible.
- Do not have monetary prizes — instead let the winner, for instance, produce a closing display for the next competition or for another event on commercial terms.
- Factor in separate funds for the fireworks themselves, labour and transport — so that competitors travelling far are not unduly penalised.

Lastly, we would love to see a competition, albeit maybe a very short one, where each competitor was given the same fireworks — so that the designer's skills could be brought to the fore. It would need a lot of organising, but would provide an interesting alternative to the excesses we often witness!

Notes

1. See http://www.tesa.org.uk/
2. See Paul Marriott's excellent webpages at http://www.montreal-fireworks.com/index.html

Chapter 16 – Pyromusicals and broadcast events

Tom Smith - Davas Ltd
Andy Wiggins - Jubilee Fireworks

The term "pyromusical" has had a chequered history. In the early days it meant nothing more than "fireworks with music", or even "music with fireworks". Today the synchronisation has reached such sophistication, mainly due to the complex firing systems available, that the combined term is truly justified. In addition of course, there has been the development of specifically modified firework types tailored to maximise the efficacy of the firing systems — in particular the use of "single-shot" devices has increased massively.

The early firework displays with music were nothing more than trying to fire fireworks at vaguely the same time as the music, aiming to start and finish together and if possible not to drown out the quiet pieces of music with noisy fireworks!

To an extent this casual approach highlights some of the problems that continue to this day. However aesthetically desirable, working with a live orchestra or band is inherently difficult. Pinning down conductors to give accurate timings of movements or entire pieces is almost impossible: despite their undoubted professionalism they are reluctant to compromise their spontaneous artistic creativity. Furthermore, conductors and band leaders are human! They may speed up or slow down by very small amounts in reaction to the audience reaction and their own enthusiasm. In a 20-minute display a 1 percent change in tempo equates to 12 seconds — this would be imperceptible to the audience but critical to the firework firer when a cake lasts, say 30 seconds, and a shell has a flight time of three to five seconds!

Add to this the vagaries of the music itself. In the famous (or infamous) "Music for the Royal Fireworks" by Handel, there is the option in the score to repeat certain sections, phrases or bars almost an infinite number of times. It is also possible to play the movements in a variety of orders (although we hope the orchestra will know the order before they start their performance!).

This makes the task of the firework display designer and practitioner in these circumstances extremely difficult.

Lastly, at such an event there is also the delay in the sound reaching the firing area from the performers. The greater the distance between the two, the longer the delay. The speed of sound is approximately $340ms^{-1}$ at 20°C at ground level, and so at 200m from the stage the delay is approximately 0.6 seconds.

Table 16.1 – Handel's firework music

Movement and broad description	Minimum time observed	Maximum time observed	Comments
Overture	4.09	9.53	
Bourée	1.02	2.03	
La Paix	2.51	3.20	Quiet and mellow
La Rejouissance	1.58	3.52	
Minuets I and II	3.11	3.30	Finales — several repetition segments

To overcome this delay it is essential that the firework firers, whether the display is fired manually or electrically, have a loudspeaker "monitor" or radio link to the stage so that this delay is minimised.

Of course, if a 100mm shell ascends 400m into the air at the same 200m distance, the delay between the visual and aural effect is approximately 1.3 seconds — so it is impossible to synchronise the visual and aural effects completely, or in time with the music. So we mostly rely on the visual effect from aerial pieces (ground noise effects can be more accurately synchronised).

All of the above, of course, leads to the conclusion that synchronising fireworks with music is only really possible when the music is prerecorded. For the highest-profile events where the presence of a live orchestra is desirable, it is often better to have the orchestra or band "mime" the piece or section to be accompanied by fireworks. The audience will, hopefully, be none the wiser!

Last of all, there is the performance of the fireworks themselves that must be considered. Fireworks are not precision explosives; they are relatively cheap, usually handmade items in which there will be some natural variation in performance.

For instance, a sample of 100mm shells tested during the development of the European Standards with exactly the same lift charges and fired from a "standard" mortar varied in burst height by up to 20 percent and burst time by up to 10 percent. The first variation is aesthetically significant, the second significant in terms of synchronisation.

The most precise devices, and hence perhaps the reason why they have been developed and used much more, are "single-shot" devices such as mines or comets where there is no perceptible delay between electric ignition and initial functioning, and the visual effect occurs immediately on ignition.

Influences of broadcast media

Increasingly, displays are designed for broadcast media (particularly live television) as well as, or even in preference to, the live audience at the site. This has meant a radical change in display design for the broadcast media and a far greater precision in timings than were ever required for live audiences alone. A modern, high-profile, broadcast-live event needs sub-second accuracy, where the broadcasters can plan (in association with the firework display designer) the best camera shots to use at any specific moment.

Thankfully, the tendency for some TV producers to superimpose firework images on top of one another and on background shots has diminished, and the true artistry of the designer can be seen in the way that it was designed.

Of course, the broadcasting of firework displays portrays a completely different view of the display to that witnessed by those watching live on the ground. Often this allows for a much broader view of the display in the context of the event and the surroundings, even if some of the drama is lost. For instance, when effects are fired from a structure it is often difficult for those close to the structure (e.g., a stadium) to appreciate the symmetry and scale of the display, especially if their view is in some way obscured.

At the Melbourne Commonwealth Games display in 2006 a new approach to integrating the display to the wider city of Melbourne was taken, and fireworks were fired from rooftops in the city and from barges along the Yarra River leading to the main stadium site. This display will be considered further in Chapter 20.

Changes in camera technology (from film-based, to low-resolution digital to high-resolution digital) and the consequent changes in the cameras' sensitivity have actually meant that the range of fireworks available to designers of broadcast events is almost unrestricted. Twenty years ago this was not the case, and gold effects, blue stars and other subtle effects simply did not show up properly and hence the palette available to the designer was significantly limited.

Synchronisation issues

Fundamentally there is the question of what is the difference between synchronisation, interpretation and choreography.

A designer could, in theory, produce an entire show that could be perfectly synchronised to a piece of music, while at the same time not interpreting the true mood and flow.

Figure 16.1 – Fireworks along the Yarra River. Photo: Howard and Sons

There is no doubt that adding a musical score to a firework display adds an extra dimension to the overall presentation and thus impression on the audience. There are arguments both for and against pyromusical displays, some based on cultural differences and expectations, some purely to do with the history and standing of the event.

The incorporation of sound and music (and possibly other effects) into the overall display design does not tell the whole story, as it is not always what you put into a display that gives the best impression; indeed there is always the valid argument that less is more. While it is true that it is impressive to see a sky full of spectacular fireworks, when designing a firework display one must be careful not to overdo things — just because you *can* do it, does not necessarily mean that you *should* do it!

Display designers today can get carried away with "cue counts". It is so easy with today's digital firing systems to pick out every single beat and syllable that one can get bogged down in over-complicated designs.

Achieving synchronisation of fireworks (either to music, or to one another) requires that the following critical information about the performance of each firework is known:

- The time from the initiation to the effect. For a mine, for instance, this may effectively be zero; for a shell it is essentially the flight time from lift to burst. If the initiating device (either an electric igniter or a manual fuse) is not directly incorporated into the lifting charge, there may be a delay between initiation and the lift.

- The functioning time of the firework itself — for instance the duration of the stars from burst to extinguishing point.

To a lesser extent the following are also important:

- The effect height — so that when a particular pattern in the sky is designed there is not undesired overlap or a gap between effects
- The effect "broadness" — for similar reasons

Once all the information is known it can be applied to a database or spreadsheet (either a bespoke programme or using standard software such as Microsoft Excel©) to determine the firing time (cue time) for each effect.

Fireworks are relatively imprecise items, however, and there will inevitably be (using current technology) some variation in flight times and burst times even for similar fireworks from the same batch. There are developments (see Chapter 18) that promise to eliminate such variations, but they are not yet widely adopted.

Chapter 17 – Electrical firing basics

Chris Pearce – Jubilee Fireworks

Introduction

"Electrical firing", whereby fireworks are ignited by an electrical signal, is not a new technique. However, in recent years there has been a very rapid advancement in the use of this method — particularly since the introduction of digital electronic systems in the 1990s. This will be discussed in the following chapter.

At its simplest level, an electrical current is used to provide a heat source of sufficient temperature to initiate blackpowder (or quickmatch). In the early days, a hot-wire was adequate — but this has now been superseded by the purpose-made "electric igniter" or "e-match" (details of which are discussed later).

Electrical methods offer the great advantage of remote firing, with enhanced safety for the display operator, although not necessarily for the spectators. Again, this consideration is dealt with elsewhere. Traditional systems were, in effect, nothing more than a large switching facility by which, at the push of a button, a voltage would be applied to a remote igniter, which in turn fired the chosen effect.

A digital system does much the same thing, but with a far greater degree of speed and control. Large and very elaborate displays can be designed using specialized software; the information is then translated into a sequence of digital commands that, in effect, fire the show. A computer can be used to adapt the showplan, even during operation, to make modifications where necessary; it is quite possible to "edit" a display in progress to make changes that affect the appearance of the show. The use of digital systems has transformed pyromusical displays to the extent that individual notes can be "hit" with a firework effect as a matter of routine, enabling the designer to produce a truly choreographed show, as opposed to something that interprets the music in a general sense.

Basic electrical theory

In order to understand how electrical firing systems function, it is necessary to study some basic electric circuit theory.

There are three important basic electrical concepts:

Current (Symbol: I, Unit: Ampere [A])

Electrical current is the flow of charged particles in a wire — it gives rise to heating in the wire. A high current can cause a wire to glow brightly (as in a bulb). Metals are good conductors of electricity — particularly copper. Current flow in a wire is represented by

Voltage (Symbol: V, Unit: Volt [V])

Voltage is applied to a circuit by a battery (or cell). If a voltage source is not present, no current will flow. A larger voltage generally gives rise to a higher current. A battery is represented in a circuit diagram by

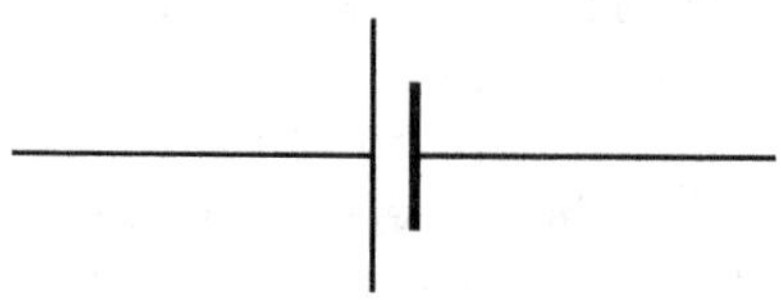

Resistance (Symbol: R, Unit: Ohm [Ω])

All conductors present a "resistance" to current flow. If only a small current flows when a high voltage is present, then the conductor has a high resistance.

A resistor (which could be any component, e.g., a bulb or igniter) is represented in a circuit diagram by

A simple electrical circuit

The simplest electrical circuit consists of a battery connected to a resistor, with a switch.

When the switch is closed, a current will flow. The current is the same at all points in the circuit. The value of the resistance is given by:

$$Resistance = Voltage/Current$$

Or

$$R = V/I$$

If a current of 1 ampere (amp) flows in the circuit when a voltage of 1 volt is applied, then the resistance is 1 ohm.

If an electrical device obeys Ohm's Law, the resistance stays constant, no matter how much current is flowing. In fact, very few devices obey Ohm's Law. Filament bulbs do not — neither do e-matches, which require a minimum current level to ignite.

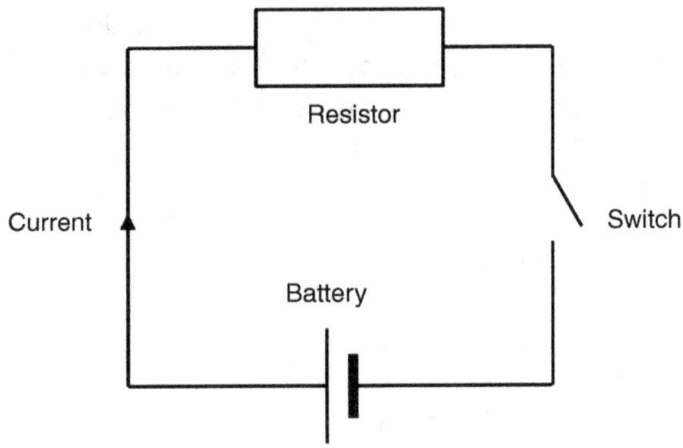

Figure 17.1 – Simple electrical circuit

The series circuit

Often, two or more devices are connected together in series across a battery. This is illustrated in Figure 15.5 for the simple case of two resistors:

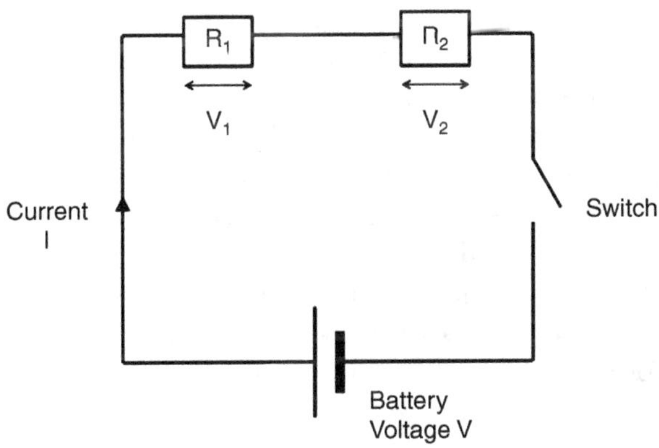

Figure 17.2 – The series circuit

In this type of circuit

- The current is the same at every point.
- The voltage provided by the battery "splits" across each resistor so that

$$\text{Battery voltage } V = V_1 + V_2$$

This can be verified experimentally by placing a voltmeter across the two resistors R_1 and R_2 independently. Voltage cannot be "lost".

The arrangement can be replaced by an "equivalent" circuit consisting of just one resistor R_T in place of the two.

So, we can say:

$$V = V_1 + V_2$$

Or

$$V = I \times R_1 + I \times R_2$$

But

$$V = I \times R_T \text{ (for the equivalent circuit)}$$

So

$$I \times R_T = I \times R_1 + I \times R_2$$

Giving

$$R_T = R_1 + R_2$$

In a series circuit, the total resistance is the sum of all the individual resistances in the circuit.

Implications for electrical firing of displays

- A simple firing system may ignite one e-match successfully, but when two or more are placed in series, they may fail. This is because the total resistance in the circuit is too high and the current is now not large enough to "fire" the igniters. One solution would be to raise the battery voltage, or use more batteries in series.
- An e-match connected directly across a 12-volt battery fires normally, but when it is connected into the firing system at the end of a long wire — it fails. This is because the wire itself has a large resistance, which becomes significant and limits the current.

Line resistance

In simple electrical circuit diagrams, we ignore the resistance of the connecting wires.

However, in practice this cannot be done, as long runs of wire cable are often used on firework display sites to connect the igniters to the firing panel (which contains the battery). The resistance presented by the connecting cable is called line resistance.

> In other words (applying the equation above),
> Total resistance = sum of component resistances + line resistance.

Worked example

Five e-matches, each of resistance 4 ohms, are connected in series along a length of cable which itself has a resistance of 16 ohms. The cable is connected to a firing panel, which produces an output voltage of 12 volts. Each e-match requires a minimum of 0.5 amps current in order to fire. Will they ignite when the panel is switched on?

Using the above equation:

Total resistance = sum of e-match resistances + line resistance.
Total resistance = $(5 \times 4) + 16 = 36$ ohms.
Now, for the complete circuit,
current flowing = voltage applied/total resistance = 12/36 = 0.333 amps.

The current flowing in a series circuit is the same for all components, so each e-match will pass 0.333 amps. This is below the minimum current required to fire them; hence they will not ignite.

Note that if just one e-match is used, total resistance = 4 + 16 = 20 ohms.
Current flowing = voltage applied/total resistance = 12/20 = 0.6 amps.
The e-match will ignite.

Disadvantages of the series circuit

If a "break" occurs at any point in a series circuit the current will no longer flow at all.

This can cause problems when "testing" a line; a very low voltage is usually applied so that the current is well below what is needed to fire the igniters, to check continuity. If no current flow is observed, it can be difficult to locate exactly where the problem is; any of the igniters could be faulty (non-conducting), or there could be a break in the circuit at any point — possibly a poor electrical connection.

It is possible for a series circuit containing several e-matches to fire only some of the igniters when a voltage is applied. If the e-matches have slightly different characteristics (e.g., resistances), they may not all ignite simultaneously. It is therefore unwise to mix e-matches produced by different manufacturers in a series circuit.

The parallel circuit

An alternative way of connecting components is the parallel method. This involves an arrangement as follows (for two components).

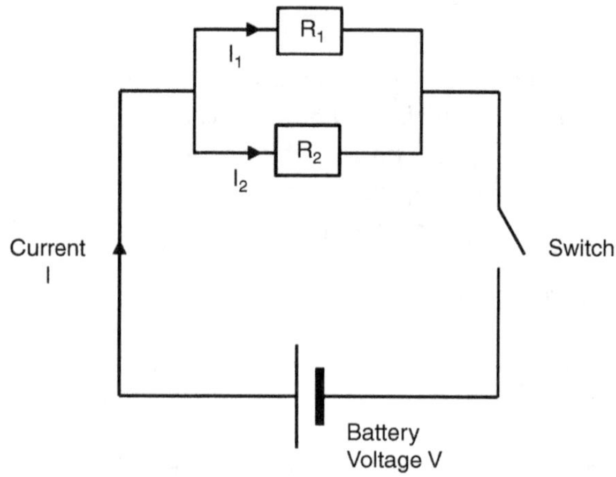

Figure 17.3 – The parallel circuit

The important features of this arrangement are

- The current effectively "splits" at the junction so that $I = I_1 + I_2$
- which means that the current flowing in each component is different (unless they have equal resistances — in which case they carry equal currents, which would be exactly half of the total circuit current I).
- The current can take different routes.
- The voltage is the same across each component R_1 and R_2.

The mathematical analysis of this circuit is more complex than in the series case; we will not consider it here.

The main practical advantage of a parallel circuit is that if one of the components becomes non-conducting, this does not stop current flow in the entire circuit. For example, if five igniters are connected in parallel across a battery (via an electrical firing panel), a fault in one would not stop the others functioning — as current can still "flow" around the circuit.

Electric igniters (e-matches)

An electrical igniter (e-match) is a pyrotechnic device used to initiate (i.e., fire) a firework remotely, using an electric current. Figures 17.6 and 17.7 illustrate a simple e-match.

Figure 17.4 – Firing controller. Photo: A DiPalma

Figure 17.5 – Firing connector box (field module)

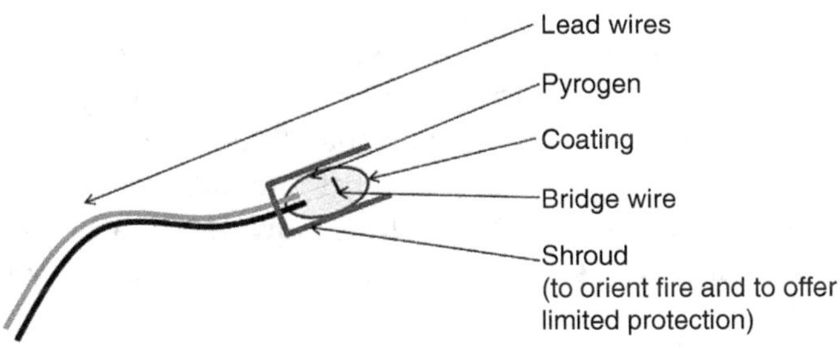

Figure 17.6 – Schematic of electric igniter

The lead wires carry the current, which passes through the bridge wire — typically made from nichrome. This heats up to "glow temperature" and fires the pyrogen (a small "pellet" of pyrotechnic compound), which burns rather like a match-head. The e-match is embedded in an appropriate part of the firework to ensure ignition — the "quickmatch" leader of a shell, for example.

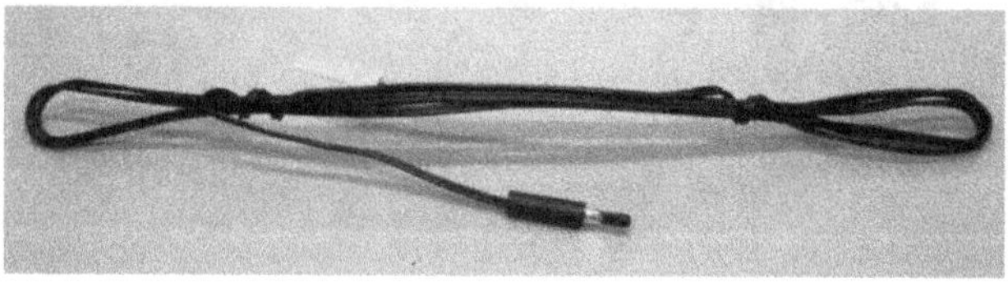

Figure 17.7 – Electric igniter (e-match)

The pyrogen material is often sensitive to both friction and impact. The shroud offers protection against accidental ignition from these sources. Many accidents have been attributed to igniters firing unexpectedly, so careful handling is important.

E-matches are essentially "pure resistors" and can be treated as such in calculations. Typical values of resistance are around 1.5 ohms. The current through the e-match must be sufficiently high to initiate the pyrogen. Commercial e-matches are rated by the "no-fire current" and the "all-fire current". The no-fire current (typically 50mA) is that for which all igniters in large batch won't fire. Conversely, the all-fire current (typically 0.1A or 1000mA) is that for which all igniters will reliably fire.

There are minor statistical variations in the resistance of igniters in any given manufactured batch, and thus their behaviour. The region between the all-fire and no-fire currents should therefore be avoided, as ignition cannot be guaranteed. The no-fire current is used for continuity test purposes in practical firing systems.

To *reliably* fire an e-match, a voltage source is applied of sufficient size to cause *at least* the all-fire current to pass through an igniter for a specified time. Currents much higher than the all-fire current still may not ignite the e-match if applied for very short intervals, as the bridge wire does not have sufficient time to reach its glow temperature. Commercial e-matches are often supplied with a table or graph describing their characteristics.

It is important to understand the behaviour and characteristics of igniters when designing an electrical circuit layout for a display or working with a particular firing system. Igniters supplied by different manufacturers will have varying characteristics, which can be significant. It is unwise, for example, to use igniters from different sources in the same series circuit.

Electrical firing systems

So far, we have considered simple circuits. In essence, traditional firing systems functioned on these principles — with outputs (called "channels") to each e-match on the firing site, via long runs of two-core wire. When a switch is closed, a voltage is applied to a specific e-match, via the firing panel, which contains a battery — typically 12 to 24 volts.

Very primitive systems would have one switch (or button) corresponding to one channel, which in turn resulted in considerable lengths of wire running across the display site. In the 1980s, more efficient systems were developed utilizing multi-core cable connections to "field modules", which would themselves provide several channel outputs to a number of e-matches. This is illustrated in Figures 17.8 and 17.9.

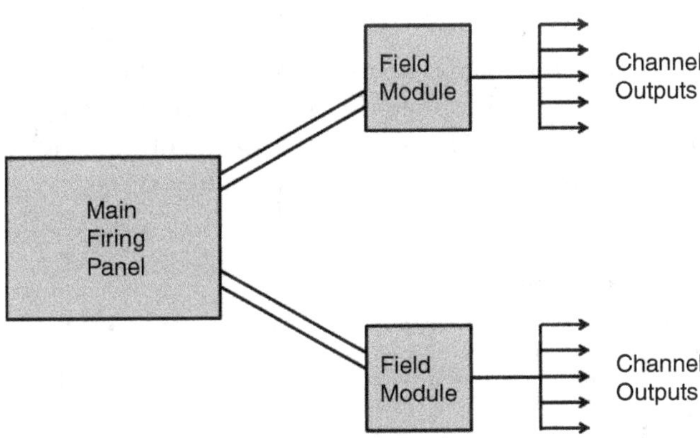

Figure 17.8 – Modular firing system

A major concern with a simple system of this type is the voltage drop that results from long lengths of cable between the voltage source in the main panel and the output from the field module. This is the problem of *line* resistance discussed earlier.

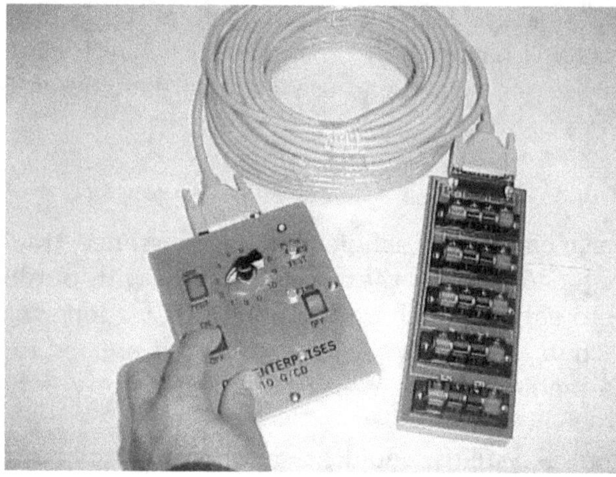

Figure 17.9 – Simple "manual" firing system with a ten-channel selector switch, multi-core cable, and a field module (with ten outputs)

Digital firing systems

This basic approach has now been superseded by systems based on digital technology. A digital signal is, in essence, a voltage signal that varies continually between one of two states — either "high" or "low" (equivalent to "on" or "off"). In digital terms, this can be interpreted as 1 or 0. A simple representation is shown in Figure 17.10, showing the difference between an "analog" signal (which varies continuously) and a digital signal, which effectively switches between two states.

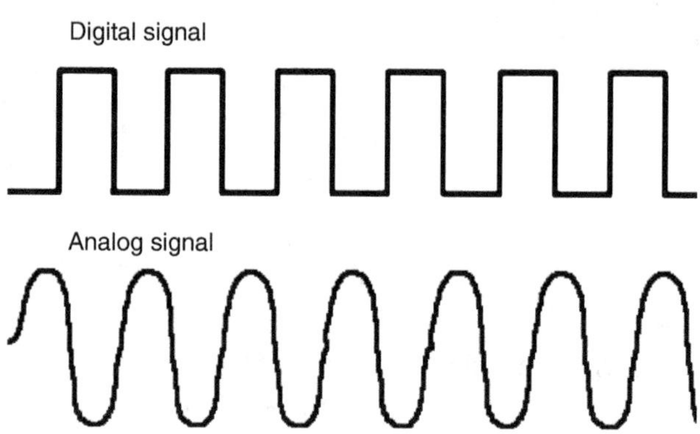

Figure 17.10 – Analog vs digital signalling

Digital systems based on computers and microprocessor technology are extremely fast; they can generate and manipulate numerous on/off states within a fraction of a second. This has enormous advantages in pyrotechnic firing systems, where individual e-matches (igniters) can be fired with incredible precision at very small time intervals. The introduction of digital systems has revolutionized the firing of professional displays, particularly pyromusicals, in the last decade.

There are numerous digital systems now available on the market. The main control panel produced by one leading manufacturer is shown in Figure 17.11.

Figure 17.11 – Firing control box

A very powerful system such as this has the following features:

- It can be used in "manual" (i.e., push-button) or full digital mode — controlled by an external laptop computer or internally downloaded file.
- It can control, test and fire up to 80 field modules (each having 32 output circuits) — giving a full capacity of 2560 (80 × 32) firing cues.
- With a laptop, the system can be used to execute a fully automatic display with the option for modifications to the show programme during firing.
- The system can also accommodate "time code" for the firing of pyromusical displays, enabling very accurate timings to be achieved.

- It is possible to expand the system easily using two or more panels.
- The system can be used "wirelessly" — using radio-frequency transmission and reception. This avoids the need for long connecting wires between the main firing panel and distant field modules — giving the advantage of being able to control displays over large distances.

The main panel, in essence, provides the digital information and power (i.e., voltage source) to the field modules. This is translated by the microprocessors in the field modules into firing signals that are applied to the e-matches attached to the module outputs. Specialised software is used to compile a "show programme" that is downloaded into the panel, which has an in-built memory. The programme is, in effect, a sequence of firing commands that are subsequently executed by the system in its automatic mode.

Some advanced digital systems have avoided the problems associated with line resistance and voltage drop — by using capacitor discharge circuits in the field modules.

The theory is beyond the scope of this book; it suffices to say that large capacitors can "store" significant amounts of electrical energy and maintain high voltages when they are "charged up" from a supply (which can be located some distance away).

Sequencers

A sequencer is basically a device in which the outputs are designed to operate in a pre-determined sequence. For example, we may wish to fire ten igniters at an interval of 0.1 seconds. Simple firework control sequencers are, essentially, devices with a single input and multiple outputs. A voltage pulse applied to the input then "triggers" the sequence, which can often be varied by selecting a particular pre-set interval. In electronic terms, this would involve changing the time intervals between the "1" and "0" states in a digital signal. Firework sequencers are useful extensions to digital (or analog) systems, as they expand the cue capability; one input signal (firing command) can initiate a large number of igniters in a set sequence.

Time code

Time code is a sequence of numeric codes generated at regular intervals by a timing system and is used for synchronisation. It is essential in complex pyromusical displays, where very accurate synchronisation is required throughout the show. The majority of digital firing systems make use of time code in some format (examples include SMPTE). A detailed discussion of the principles of time code is beyond the scope of this book — but users of digital firing systems will invariably come across its application (see below).

Recent advances have included GPS synchronisation — whereby the firing system is fitted with a GPS antenna and is able to lock into the *Global Positioning Satellite* facility. This enables the firing system to locate its position, very accurately, anywhere on the earth along with a definitive "time".

Basic principles of software applications

When planning a pyromusical display, there are five typical steps in the process of show design — which is undertaken using software provided by the system manufacturer. Software "packages" vary, but the following procedure is typical.

Create the musical score – The Musical Score software enables the designer to edit the chosen music into a "score"; often the designer will want to combine several pieces. The software usually provides a visual graphical representation of the music on a volume-vs-time scale.

Choreograph specific fireworks to the music – Particular fireworks (e.g., shells, Roman candles, single shots) are selected to integrate into and "interpret" the music programme. This software maintains a database inventory of products, and a particular effect is selected to be "dropped" into the music at a desired time — thereby creating the choreographed display.

It is usually possible to adjust timings to allow for natural delays — for example, the rise time of a shell. The "burst" (which may be required to hit a particular note in the music) will occur several seconds after the mortar launch.

Assign choreographed product to locations and cues – Having designed the show, the designer must now assign each product (and time) to a particular "cue". This is, in effect, an e-match located at a particular point on the display site and thus a given channel on a specific field module (e.g., 100mm red chrysanthemum shell at time 1.05 = module 4 channel 6).

A specific software package enables this, which in turn enables the firing system to actually fire the display as required.

Create time code track to synchronize the music to the fireworks – TimeCode software creates perfect digital time code from the music score created in Step 1. The time code is necessary to synchronize the choreographed fireworks to the music score. This software creates a perfect digital audio recording of the music, with the digital time code applied on a separate channel.

Fire the display! – The digital firing of the display is then executed by another piece of software in conjunction with the hardware (i.e., the main control panel, field modules etc.). This particular software often provides setup and testing routines to verify proper operation of the system prior to firing the display (for example, e-match integrity and continuity). Additionally, some system software may be used to download the display into the control panel for firing without a laptop computer. Where a laptop is used in conjunction with the main panel, the executed show programme can be monitored as the display progresses, and in certain cases, modifications can be made (removal of certain products or cues, for example).

Chapter 18 – Firing systems review

Rodney S J Clarke – FireOne

Firing systems, whether traditional manual or modern digital, are among the most important tools available to pyrotechnicians, enabling the safe delivery of choreographed displays varying in size from the smallest to the largest imaginable. Although hand firing is still a popular pursuit for some, every modern firework display company will employ a firing system of some sort to deliver the majority of its shows throughout the year. Firing systems come in all manner of shapes, sizes and capabilities; however, there are several fundamental principles that govern their overall design and operation. This chapter will outline the natural development of firing systems, from the simplest to the most complex.

As mentioned previously, the most basic electrical firing system for pyrotechnic devices consists of an e-match, lengths of connecting wire, a DC power source and a switch mechanism to close the circuit, thus controlling the precise moment of e-match ignition. This arrangement is well demonstrated both visually and technically by the forefather of all modern firing systems, the nail board, which is still in service to this day due to its sheer simplicity and durability.

A simple nail board (Figure 18.1) normally comprises a wooden plank with multiple nails driven through the timber at regular intervals in a single long line. Each nail is wired to one side of an e-match, with the other side of the e-match either wired to the other nail or connected (or commoned together in the case of multiple e-matches — see Figure 18.2) to one terminal of a direct current power source, i.e., a battery. Another piece of wire is then attached to the other terminal on the battery and the opposite bare end is wound around yet another nail. When this free nail is touched to any other nail on the board, the circuit is closed, electrical current flows and the e-match ignites. Running the free nail along a series of nails in the board creates a rapid firing sequence, a technique that has been used many times over the years by the film and television industry to simulate bullet hits and strafes in battle scenes and war films.

A natural progression of the simple nail board is to replace each nail in the board with a momentary push-button switch. Another alternative is to replace the multiple nails with a multi-position rotary switch and the free nail with a momentary push-button switch. Rapid-fire sequences can still be achieved with this latter arrangement simply by holding down the push button and then turning the rotary switch through

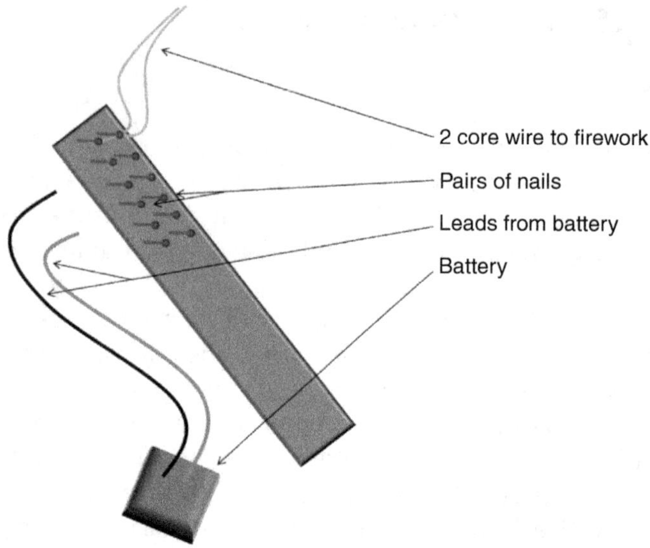

Figure 18.1 – Schematic simple nail board

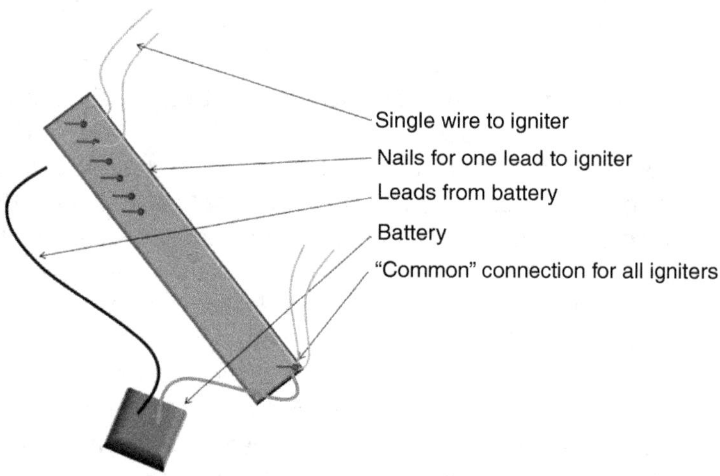

Figure 18.2 – Nail board with "commoned" terminal

multiple positions as required. In most cases, multiple lengths of two-core bell-wire used to connect e-matches to such basic firing systems are replaced with a single reusable multi-core cable terminated with multi-pole connectors. The far end of the multi-core cable is then frequently terminated in a "rail" or "slat", which normally presents the two sides of the circuit as a basic push to release connectors such as those commonly used to connect low-resistance wire to audio speakers.

A further development of these principles uses multiple push-button switches as "fire" buttons combined with a rotary switch or similar to progress through multiple rails or slats. This arrangement is achieved via multiple ground or common wires in addition to the multiple "live" wires. The benefit of such an arrangement is that the maximum number of "cues", i.e., individual pyrotechnic products that can be fired separately, increases significantly without drastically multiplying component cost. In the case of a firing system with ten push-button switches and a single ten-position rotary for switching commons, the resulting cue count is 100 using a cable with 20 cores. As the wiring complexity of a firing system increases, so do some of the electrical considerations. In particular, switched common systems must utilise diodes to protect the commons from providing alternative current paths that would otherwise result in misfires.

One very useful feature missing from these manual systems is the ability to safely test the firing circuit without prematurely firing the e-match, thus providing the operator with an indication that everything has been connected correctly. Commercial e-matches will have a stated no-fire current, which is the maximum electrical current that can pass through the e-match without it actually igniting. By including a suitably rated resistor and a Light Emitting Diode (LED) in the circuit, it is possible to provide a clear visual indication of whether the firing circuit is open or closed. As it is always a remote possibility that an e-match may fire at less than the no-fire current, a display site MUST always be cleared of all personnel before any testing is conducted. Furthermore, it is both practical and prudent for a test circuit to utilise a fraction of the specified no-fire current. In comparison, most mainstream commercial e-matches have a no-fire current between 200 and 300mA, whereas most LEDs require only a few tens of milliamps to illuminate brightly, thus providing a significant safety margin.

Manual firing systems have been used very effectively in the firework display industry for several decades; however, they suffer from one key design limitation that could also be considered a potential risk to safe system operation. That flaw is the human operator, who needs to press the right buttons at exactly the right time. Failure to do so may result in artistic failure at best, or a dangerous conflagration at worst. Furthermore, a human operator cannot achieve precise and rapid non-sequential firing, which in turn limits the creative delivery of the design director. As the electronics industry underwent rapid advances in the latter quarter of the last century, it was not long before someone removed human operators from the firing equation and replaced them with semiconductors instead.

In the late 1980s, several groups started to consider the possibilities of truly digital firing systems. It took several years of research and design before the first relatively crude systems came to market. These systems replaced the paper firing orders and stopwatch with a digital script that could either run from a personal computer or be downloaded into the bespoke firing-system hardware for later recall in the field. These first tentative steps into the world of digital firing systems revolutionised what could be delivered artistically. Within a single decade, a multitude of competing systems were developed, thus creating a completely new market within the pyrotechnics industry.

Figure 18.3 – Firing system controller

Most digital firing systems adhere to a common hardware architecture and hierarchy. There is at least one controller or panel (Figure 18.3), which relays instructions to modules placed amongst the pyrotechnics out in the field. These modules are coupled to rails or slats to which the actual e-matches are connected. Very often the module and rail are combined into a single unit (Figure 18.4), although when separate permit a broad variety of rails to be used to best suit different requirements.

Long racks of mortar tubes benefit from long connection rails running alongside the shell block, whereas large concentrations of single shots are better served by a more compact connection rail. The cue count for each module/rail combination varies significantly from system to system; however, values between 16 and 32 are very common. These field modules are then connected back to the main panel via control cables or even wireless transceivers (Figure 18.5 and Figure 18.6).

Figure 18.4 – Field module and rail

One key benefit of digital firing systems is delivered by the distributed processing undertaken by the field modules. These "black boxes" can do far more than just serve as a means of being able to connect e-matches back to the controller; they decode potentially complex instructions sent by the panel and then act accordingly. Using digital signalling removes the requirement for expensive multi-core connectors and multi-core cable. Most modern systems utilise nothing more elaborate than two-core cable, which can be obtained at a fraction of the cost of cable used in traditional manual systems. Most systems also use relatively cheap XLR connectors to connect the various hardware components to each other, and some systems have actually done away with the connector altogether by using bared ends of the cable inserted into spring-loaded connectors on the controller and module. This simplicity of connection belies the complexity of data transmission taking place behind the scenes.

Figure 18.5 – Wireless transceiver

Sometimes it is neither practical, nor feasible, to run command cable from where the pyrotechnics are located back to the firing position, e.g., in multi-barge or multi-rooftop displays. In such circumstances wireless transceivers (a portmanteau of transmitter and receiver) can be pressed into service to replace the usual cable "home run". Wireless transceivers come in various power levels and in some cases can broadcast data over an area of several square miles. Transceivers commonly use frequency-hopping spread spectrum technology and encryption algorithms to ensure that firing-system data cannot be hijacked or altered. Wireless connectivity is often viewed with critical suspicion by some experienced pyrotechnicians; however, the technology has matured very quickly and proved itself viable in environments subjected to considerable radio interference. Ideally, any wireless technology should be implemented in such a manner that its operation is transparent to the user, and that no special or convoluted additional procedures are required for it to function successfully.

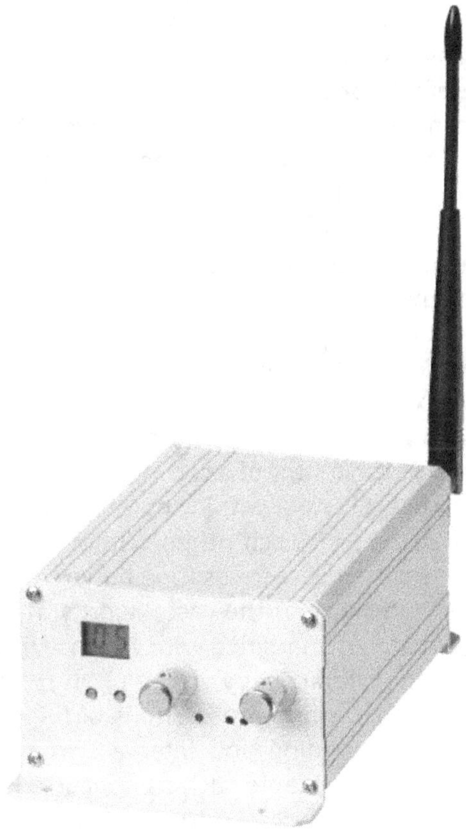

Figure 18.6 – Wireless field module

Along with distributed processing, digital firing systems face a key design decision: the question of power distribution. With traditional systems, all power is provided by the controller and therefore long cable length restricts total firing potential. There are three key power options available with digital systems. The first is to use the main panel for any direct power requirements, which has similar constraints to traditional systems. The second is to implement a local power source in each field module, i.e., a battery; however, that incurs a maintenance and support overhead. The third is to use capacitive discharge technology whereby modules store up firing potential delivered by the panel. This last option is an effective compromise as only one power source needs to be charged and maintained, and sufficient firepower is available close to the point of ignition even when using very long lengths of command cable. Sometimes a combination of techniques is required, especially when wireless transceivers are used and a power connection from the main panel is therefore not possible.

Regardless of connectivity type, all digital firing systems are dependent upon a custom command protocol in the form of electrical signalling to actually implement all the various testing and firing functionality. These command protocols are very much proprietary and confidential for each manufacturer and therefore vary significantly. Probably the most important element served by the hardware design combined with the command protocol is that pyrotechnic product can only be fired once several safety interlocks have been achieved. Command protocols need to be robust, extensible, invisible to the user, and above all, safe.

The first digital firing systems to appear on the market delivered a firing resolution measured in tenths of a second; that is, two cues could be fired independently of each other within approximately 100 milliseconds (one tenth of one second) of each other. Within a few years this had improved down to just 10 milliseconds (one hundredth of one second), which is more than sufficient for the vast majority of applications. The response time of e-matches combined with variance in the burn time of pyrotechnic compositions eradicates any value in firing-system manufacturers striving to achieve firing resolutions of less than 10 milliseconds. However, firing resolution and intervals do not represent the whole picture. Advanced concepts such as concurrency, accuracy and precision also have a major part to play in the design of any digital firing system.

A common requirement for many displays is the simultaneous lift of the same effect from multiple positions along a frontage. One way to achieve this effect would be to wire each of the effects into the same firing circuit; however, this has several disadvantages. Irrespective of the additional work effort required, depending upon the wiring method used, either a single faulty e-match could prevent the entire effect from firing, or a high firing current would be required to successfully fire all positions. Another alternative would be to script the firing of each position at the same time using multiple field modules. This latter option, although elegant, can potentially highlight a limitation of many digital firing systems, as the firing command for each cue is sent by the panel consecutively rather than concurrently. Even when the time taken to send each firing command is measured at around 10 milliseconds, the human eye can discern a visual difference when the position count enters double figures and what should be a simultaneous lift instead appears as a very fast chase sequence. When the number of positions involved in a simultaneous lift exceeds 30 or 40, then the communication lag inherent in most digital systems becomes very obvious indeed.

One solution to this problem is to leverage the processing power of field modules in much the same way as the conductor of an orchestra can control a large number of musicians simultaneously. Rather than have a panel send out individual firing instructions to each module, each module can be downloaded with the relevant portion of the firing script and then the panel send out regular time messages to which the modules respond by firing their portion of the script. In this way, a very large number of modules will fire product absolutely simultaneously. There are very few systems that have this advanced capability, which has been used to produce some of the most spectacular effects yet envisaged. Obviously it is very important that such a capability incorporate both internal and external safety protocols so that it is impossible for an individual module to act autonomously outside the control of the main panel. Although concurrency is a very powerful feature, firing systems must also be very accurate and precise in the context that cues must be fired exactly and repeatedly on time. Systems with low accuracy/precision can easily ruin an otherwise carefully choreographed effect due to high variance in the firing window.

Key to the operation of all digital firing systems is an accurate time source. Sometimes this is delivered internally by a timer built into the panel, but very often the firing of a show has to be tightly synchronised with other elements such as audio and lighting. This requires an external means of coordinating and controlling all the various systems, which is often achieved via a technology borrowed from the motion picture and television industry called timecode. Timecode is simply an absolute digital time mark encoded as an audio signal that can be interpreted by compatible equipment. A very popular type of timecode is called SMPTE (Society of Motion Picture and Television Engineers); however, this is not particularly well suited to the harsh operational environments sometimes imposed by the fireworks industry, so a more robust alternative called FSK (Frequency Shift Keying) is also frequently used. The main advantage of an FSK timecode signal is that it can be transmitted over long distances using comparatively low-fidelity devices such as two-way radios.

More recently, some firing systems have also been equipped with GPS (Global Positioning System) decoders, identical to those used in consumer satellite navigation systems, as GPS signals contain very accurate time data and are available anywhere in the world. Via GPS, multiple firing systems can operate totally independently, but still be in perfect synchronisation with each other without having to implement yet another centralised system just to broadcast timecode. Firing a show using GPS as a time source can be as simple as setting the current time zone in the panel, specifying a start time for the show, plugging in the GPS antenna, and then arming the system.

With the growth in the capability of firing-system hardware and the high degree of scalability now available, the cue count routinely used in firework displays has also steadily increased. Shows now regularly contain between many hundred and several thousand cues, and that in turn has required advances in how displays are initially designed and prepared. Software applications have been developed to specifically cater to the needs of the fireworks industry, allowing designers to maintain extensive product databases and quickly incorporate their inventory into display scripts. This design software is invaluable for the production of pyromusicals, as the application can automatically implement any necessary back-timing so that effects are launched and appear in the night sky as and when required in synchronisation with the music. Some applications can even provide a 3D visualisation of the display on-screen so that designers can see whether their creative vision will match reality. Such tools are also very useful during the tendering process, as a display company can provide the prospective customer with something tangible and visual that has been tailored to their specific requirements.

Along with design software, there has also been an opportunity to develop field-testing and firing software that far exceeds the capabilities delivered by hardware alone. Ensuring that a show will fire as required is an integral part of the display setup process. Software applications can now identify issues with the system from the hardware panel right through to the actual e-match used to ignite the product. This same software can then be used to manipulate the display in real time as it fires (Figure 18.7). Timelines can be shifted to speed up or slow down the rate of fire, firing positions can be activated and deactivated as necessary, even specific product types can be withdrawn from a show in the event that it may present a hazard if it were to fire. Modern PCs coupled with modern bespoke firing hardware present a very powerful solution for delivering complex and artistic displays safely.

Digital firing systems have certainly developed and changed significantly over the last 20 years with a wide variety of new technologies

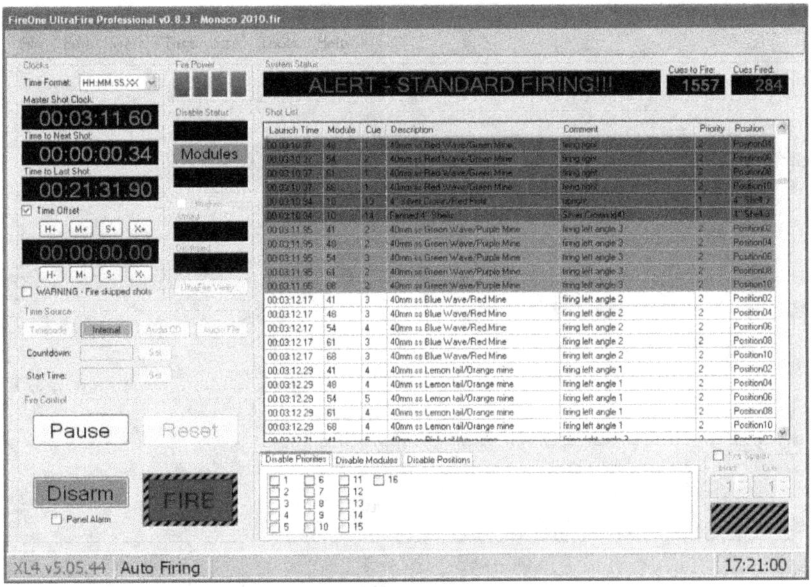

Figure 18.7 – Field testing and firing software

being exploited to deliver new capabilities and functionality. Although the pace of change has slowed as systems have delivered what has been most demanded by the industry, innovation is still taking place as new requirements are being dreamt up by pyrotechnic “imagineers” around the globe.

Chapter 19 – Case studies — small- and medium-scale events

Tom Smith – Davas Ltd
Martin Smith – Brockham Bonfire Ltd

Introduction

In the site diagrams that follow in this and the next chapter, the following colour coding is used:

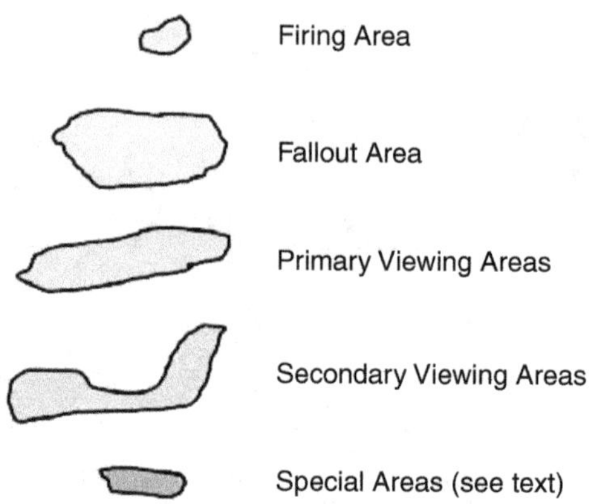

Figure 19.1 – Key for Google Earth diagrams in the following sections

Just because an event is private rather than public, it doesn't mean the event is necessarily small, nor is it always simple to set up and fire. Private events range from the amateur firing of a small display for a wedding to quite complex professionally fired displays in restricted areas for private functions.

This book does not address the firing of theatrical items indoors or as part of a larger outside display — although many of the same principles apply.

Local events (e.g., a display for a primary school)

Small-scale events for local schools are typical of the type of event fired in thousands of locations in the UK every November. As outlined in Chapter 4 there are opportunities and constraints that have to be considered.

Figure 19.2 – Location and features of small school display

Local events such as this are an increasing feature of the traditional "Gunpowder Plot" celebrations, particularly as the use of fireworks at home is either discouraged or becomes prohibitively expensive. Small community groups, such as schools, stage a small display — either fired by amateurs using category 3 fireworks, or by the smaller professional companies (very often the large professional companies are too busy or too costly at this time of year).

It would be quite wrong, however, for governments in any country to try and ban the use of suitable fireworks by individuals. Provided that the fireworks are of authenticated quality — and critically, that the user actually reads and obeys the instructions provided — the risks from such items are low in comparison to the number sold and used.

The site

The site shown in Figure 19.2 is typical of these small displays. It consists of a school playing field that is both the firing area and the fallout area, and a relatively small area for spectators that probably also features provision of food, drinks and other consumables.

The total area is not great, but the firing area is positioned so as to maximise the fallout area under a variety of conditions. If in the case shown the wind were northerly, then the audience could be moved to the far west of the sports fields.

The choice of fireworks

The fireworks available depend on who is firing the display. A comparison is given below.

Type	Cat 3	Cat 4	Comments
Rockets	Yes	Yes	Rockets are infrequently used in cat 4 displays because of the lack of precision and debris problems.
Fountains	Yes	Yes	Cat 4 displays often fire fountains from many places simultaneously. The fountain duration many also be significantly reduced (e.g., a "jet").
Mines	Yes	Yes	Cat 4 displays often use mines fired from several places simultaneously (a front) or sequentially (a chase).
Set-pieces	Yes	Yes	Traditional set-pieces (wheels, static pieces), which add duration to the event and reduce the intensity
Multi-shot batteries	Yes	Yes	Cat 4 cakes may be more varied in style and effect than their Cat 3 equivalents.
Lancework	Yes	Yes	Expensive but range from "GOOD NIGHT" to a sponsor's logo and to "battles" or novelty devices, which appeal particularly to the younger age groups
Shells	No	Yes	Shells are not available to amateur firers in the UK. Their use has largely been supplanted by the use of multi-shot batteries and by rockets.

As the display is for a primary school (pupil ages from say 4–11), there are quite varied demands on the intensity of the display. We have found that it is essential that the intensity and noise levels of the display are increased during the show, so that the display does not start too dramatically and upset the audience and have them leaving before it is finished (and the opportunity for raising money is not yet complete).

Setting up

The fireworks are usually tied to stakes driven into the ground, or in the case of larger multi-shot batteries, are placed onto the ground directly. In general, "digging in" of even the smallest items is highly frowned upon on school playing fields and care must be taken after the display to ensure there are no fragments left that could cause injury to the children. Usually a few good mowings will pick up any remaining pieces.

Firing the display – methods

Small displays like this are very often fired manually and in rotation (see Chapter 11) in order to keep continuity and avoid any pauses that could lead to the young audience becoming restless. As noted above, it is often very desirable to start the display "small" and build in intensity and volume as the display progresses.

Amateur firers should thoroughly familiarise themselves with the lay-out of the site and the specific instructions for each firework or firework type well before the display. In general, however, such displays should be fired by a maximum of three people; any more and the communication between firers is likely to become confused and ambiguous! As the firers are often parents or members of the Parent Teachers Association, there will normally be plenty of other jobs for willing volunteers to do!

Figure 19.3 – Multi-shot battery at small school display

Brockham bonfire

Brockham Bonfire[1] is a traditional village event held on or around November 5th to commemorate and "celebrate" the failure of Guy

Fawkes to blow up the Houses of Parliament in 1605; it is held in Brockham Green, Surrey, about 30 miles south of London. The celebrations in their current form can be traced back to the early 1930s, with a large bonfire constructed of brushwood and tree cuttings, and a firework display. It is lit by torches carried around the village by a traditional procession. The bonfire used to be close to 50 feet tall but has been reduced to 25 feet on safety grounds. This smaller bonfire still takes over six weeks to build. In recent decades the event has gained popularity (in part as a result of other local celebrations being cancelled on the back of onerous insurance and health and safety restrictions). Today over 20,000 people attend, making it one of the largest Guy Fawkes celebrations in the country. While there is no entrance fee, significant sums are raised each year and distributed to local charities and good causes.

The site

The whole event centres on the small and picturesque village green. The bonfire is built in the middle of the green, and a large cordon of chestnut fencing is constructed around the bonfire. As the village green is surrounded by houses, shops, pubs and the village church, space for the spectators is limited. Access points both for normal movements of people and potential emergency situations are also restricted, and significant effort (including a pre-event tabletop exercise with the local emergency services) is necessary to plan for all eventualities.

Figure 19.4 – Location and features of Brockham Bonfire

Figure 19.5 – The bonfire

Firing the firework display

Up until the early 1980s, the fireworks were fired from a sectioned-off site in the middle of the village green. The display consisted of numerous set-pieces including "pigeons" that flew over the heads of the spectators and relatively small candles, Roman candles and shells up to 6". The spectators were very near the fireworks and there were several, though minor, incidents of spectators being burnt by firework debris. With the whole event becoming increasingly popular and with safety considerations becoming ever more important, a decision was taken to move the display to a field north of the village green where it continues to be fired today. In part this change was necessitated by the increasing popularity of the event and the need to maximise the viewing area for the public, but also by changes in display design and a desire to use a wider variety of fireworks.

The modern display

Moving the firework display off the village green had a number of implications. Most importantly, the site is much safer. There is a physical barrier of a row of houses between the firing field and the spectators with a minimum separation distance of over 100m. The firing field also has a single entry point and is bordered by the river Mole, making it relatively easy to ensure it is secure — security guards are posted at the field entrance before and during the display. Residents whose gardens back onto the firing field are encouraged to stay as far away from the field as possible.

Figure 19.6 – Mortar racking — note modern "open ladder" design

The other big change is the type of display. Clearly the vast bulk of the fireworks now have to be aerial, and given the height of the houses we do have to ensure that all the fireworks chosen rise to a sufficient height to be seen by the spectators on the village green. For a number of years we did have a set-piece made of lancework positioned in the front garden of one of the houses. However, this was never quite satisfactory, with smoke as well as trees stopping it from being seen by a large part of the crowd, and we no longer do this.

Figure 19.7 – Finale racks and Roman candle batteries

Figure 19.8 – Overall view of the firing site

This leaves the current display made up entirely of shells, mines, Roman candles and large-calibre cakes, and we do have to make sure the cakes and Roman candles in particular are of sufficient size to rise high enough to be seen. One of the advantages of the site is that we can spread it across the length of the field. We set up three main positions for firing the shells, with up to a dozen positions for Romans and six positions for cakes stretching for over 100m. This gives a very full effect for the spectators watching from the village green.

We did fire the whole display electronically using Pyromate™, but this involved running very large distances of wire to the most distant cakes and Romans. In one year we used 15 kilometres of wire. We now split the display between Pyromate (for most of the shells) and Firelite™ for the Romans, cakes and the finale, giving us greater control and significantly reducing the amount of wire needed. Each year we buy a few more modules, and we hope to fire the entire display using Firelite in the next few years. Firelite has given us new opportunities for firing chase sequences and other variations on the standard sequences. We of course pay particular attention to the finale.

Specific issues

Our biggest problem is the weather, especially as the site is within the floodplain of the river Mole. A few years ago, the entire firework field flooded the day after the display. We managed to remove all the valuable equipment, but piles of spent cakes, scaffolding poles and stakes

Figure 19.9 – Shells and splitting comet Roman candles as seen from centre of the "Green"

Figure 19.10 – Shells, bombette and comet Roman candles as seen from centre of "The Green"

remained in the field for weeks afterwards until it was dry enough to get vehicles back in. We do have an emergency alternative firing site that is slightly higher just in case, but have never had to use it yet. The biggest ongoing issue is the damp, and we have to be scrupulous in protecting all the wiring to avoid any failures.

Figure 19.11 – Bonfire burning some six hours after being lit

Many of these problems are not unique to this display of course, and a considerable effort is generally needed to protect fireworks from the effects of the weather as well as from accidental ignition from adjacent items. This problem is exacerbated because of the extended periods for which the fireworks are in position — in some ways the old displays, where items were "issued" from temporary stores and loaded immediately prior to firing, removed such issues!

Brockham Bonfire has continued to develop their display to meet the challenges of modern concerns and of audience expectations.

Figure 19.12 – Splitting comet shells at Brockham Bonfire

Concerts

The demands of a small outdoor theatrical event and concert provide an opportunity to examine the intermittent firing of fireworks rather than a continuous display.

Claremont Landscape Garden[2] is a National Trust property to the south of London and was designed and developed by many of the great names in garden history, including Sir John Vanbrugh, Charles Bridgeman, William Kent and "Capability" Brown. Part of the garden is set around a lake and island, and it is from the island that fireworks were fired in a series of concerts to be watched by the audience on the shores and in the amphitheatre clearly visible in the image below.

The events at Claremont (and at many other properties owned and managed by the National Trust) are varied in scope and nature, but typically involve the intermittent use of fireworks during the performance held on the main stage and culminating in an extensive display at the end of the evening.

Figure 19.13 – Location and layout of Claremont Landscape Garden event

Site layout

The fireworks are set up on the island — which necessitates moving them to the island in the first place! The island is, however, a secure venue with no problems of public access.

One particular issue with working on small islands such as this is the small depth of soil before the water table is reached — and fireworks that need digging in should be adequately protected from the possible ingress of water. The remainder of the fireworks are usually affixed to stakes driven into the ground or are self-supporting.

The lake provides a natural barrier and potential fallout area for the fireworks, but one should not imagine that all the debris will necessarily fall harmlessly into the lake. Angling the fireworks back from the audience and choice of suitable fireworks is extremely important in what is a relatively tight site.

Timing and communication

The intermittent use of fireworks demands good, reliable and clear communication between the event producers and the firing crew on-site. Unless the display is of sufficient size and complexity to necessitate all-electric firing (which can then be controlled near to the stage), the various cues must be unambiguous and the effects chosen carefully to complement but not overwhelm the action on the stage. Prior planning will have determined the expected duration of each section of the stage action — but do not forget that live performances rarely run exactly to time!

Typical content

The main display is based around sequences of Roman candles and shells, supplemented by the use of water fireworks and fountains to mirror the quiet sections of the music. Because of the prevailing wind direction (SW) and the relative proximity of the audience (c. 200m), the maximum calibre of shells chosen was restricted to 100mm.

Issues

The main issue with events of this scale is how to maintain an adequate concentration of fireworks in the sky to satisfy the paying audience, without overwhelming the action on the stage (assuming that the budget allows for such a problem to exist in the first place!). The subtle use of fireworks, often fired at low angles to be visible at either side of the stage while keeping back the main aerial display for the finale when the action on the stage is complete, is one way of achieving this at Claremont — but there are plenty of other ways!

Plymouth – British Fireworks Championships

One of the unique features of the Plymouth Competition, the principles of which were described in Chapter 15, is TESA's encouraging of other UK display companies to come and visit the display site. Over time this has led to a very significant overall improvement of the display quality — both in terms of the final display but also in the way people set up and fire the show.

Many experienced visitors to the firing site will see ten things in the displays, decide they could do better in nine of them, but see one technique, or clever use of materials, or firing pattern, that makes them reassess their own operations.

The display site and rigging the displays

The display site is a solid surface, the Breakwater, on which each competitor has a firing area of approximately 30m × 7m to set up the display. A gap is left between each competitor, and an access route is left at the rear of the breakwater for emergency or other access. All structures used must be essentially self-supporting, although the competitors are supplied with sandbags and it is possible to affix some pieces to the railing at the font of the display site.

Figure 19.14 – Location and layout of Plymouth fireworks competition

Six displays are fired, three on each night, and so in the 12 years in which the author has been involved as adjudicator of the competition there have been some 60 different companies (some companies have fired more than once) firing, and there has only been one occasion when we have had to require a competitor to put additional measures in place to ensure the safety of the display.

Companies differ quite widely in their setup techniques and the extent to which special equipment has been developed for the particular site. Although the firing area allocated to each competitor is relatively small, most competitors look to increase the "width" of the display by firing Roman candles and smaller shells at significant angles.

The viewing areas

The main viewing area, and the area used by the judges, is situated on the Plymouth Hoe and the area around the Citadel — some 400m from the firing site across the main Plymouth harbour. This elevated position is just about perfect — it is high enough that the reflections in the water enhance the display while still giving a good view of all the various effects used. It is not too high, or so far away from the firing point that the impact is lost, and of course, it provides an extensive area to accommodate the audience safely.

Figure 19.15 – Mortar racking at Plymouth

The other main viewing areas are on the cliff to the south of the firing site (although this is a long way off) and from vessels moored in the harbour itself — indeed the lights from these vessels add to the spectacle.

Figure 19.16 – Large-calibre mortars at Plymouth

Firing the displays

Depending on the weather and constraints from the broadcasters, the first show is fired at c. 21:50. Each show lasts ten minutes and there is a ten minute gap between the displays — giving some 50 minutes of entertainment overall. We find a ten-minute gap is ideal — it allows a clear distinction to be made between each competitor and for a broadcast introduction (if it is to be used) to be made without prolonging the pause between displays to the point that the audience becomes restless and cold.

If the weather is poor, the gap between the displays is reduced. Each competitor is in communication with the adjudicators and is cued in by them — the timing for each display is from the last of three signal maroons to announce the display, to the last of three to signal the end. Although this is somewhat old-fashioned (it used to be a regular feature of major displays in the UK) it gives an accurate timing point and gives the audience and broadcasters a positive marker to determine when each show starts (or to make people look in the right direction to see the true start) and the finish.

Figure 19.17 – General layout of Mountbatten Breakwater

Issues

The display site is well removed from the public and is easily controlled — this is what makes it so ideal — but even then there have been problems.

In one of the first years of the competition an unexploded bomb was found adjacent to the breakwater — this caused the postponement of the first night's shows and the firing of six on the second night — which caused significant issues of space. Thankfully this has not been a regular feature of the displays!

The breakwater is on a highly tidal estuary, and if the wind is from the south waves can (and do) break over the breakwater. Thankfully this has only been an issue once (the weather in mid-August in the UK is normally pretty fair).

What is certain around most of the world, however, is that if it is high water at 07:00 in the morning then it is likely to be high water at 19:30 in the evening. If the waves are breaking over the breakwater when the competitors begin their setup, there is a distinct possibility that the fireworks will be at risk again just prior to the displays. It never ceases to amaze how few people understand tides!

In the early years of the display each competitor had their fireworks weighed before setting up could take place. It is incredibly difficult to try and ensure fairness in competitions (see Chapter 15), and we decided that the weighing process caused unnecessary delays and threatened the smooth running of the overall event. As explained above, each competitor is required to provide a full breakdown of the items to be used, and this is checked on-site by random sampling.

Water fireworks were used extensively in the early displays, but the use has declined somewhat in recent years. Part of this is the realisation by competitors that they are actually a long way from the audience. Competitors are encouraged to gauge the height of a person standing in the viewing areas on the Plymouth Hoe and understand that a 2m effect on the water (like a water mine containing small fountains or strobes)

Figure 19.18 – Jubilee Fireworks at Plymouth. Photo: Jubilee Fireworks

will look as big to the audience as a person looks to someone on the breakwater! We have also had significant problems with the safety of water fireworks in the past, and this has led to restrictions being placed on their use. The major reason for this is simply the lack of data available to the user about the consequences of changing mortar length, calibre and exact firing angle of water fireworks.

Lastly, general fallout from the display can be a significant issue even though the site is well isolated and extensive. In a harbour area the wind can often swirl and be funnelled around and between the masses of land on either side of the firing area, and "normal" debris and long-burning stars can travel significant distances downwind — and the wind direction can be quite different at the bursting height and near the water level.

Types of fireworks used

The competitors are not unduly restricted in their choice of fireworks (with some constraints), and the fireworks used in previous displays include:

- Shells — ranging from 50mm to 300mm (300mm is the maximum allowed, and the use of these shells may be restricted if the conditions dictate)
- Rockets — usually flight rockets
- Roman candles — ranging from 18mm to 75mm calibre
- Mines — from 75mm to 150mm calibre
- Multi-shot batteries — every possible combination of calibre and effect
- Single-shot devices — typically 30mm and 45mm comets fired in "chases" — see Chapter 11
- Set-pieces — a number of competitors have produced custom-built wheels and set-pieces for their displays — the largest was some 10m in diameter
- Fountains — up to 75mm calibre — usually fired along the available frontage of the breakwater and also horizontally over the water
- Flying saucers (aerial wheels), which can look very effective fired over the water, but their use may be restricted if the conditions dictate
- Water fireworks — shells, bombettes etc.

Notes

1. See http://www.brockhambonfire.com
2. See http://www.nationaltrust.org.uk/main/w-claremontlandscapegarden

Chapter 20 – Case studies — very large-scale events

Tom Smith - Davas Ltd
Andrew Walsh - Accolade Events
Jim Donald - Jack Morton Public Events

The planning of very large-scale events is a highly complex process, done over months or even years and involving a multitude of interested parties. These high-profile occasions, while often not generating large profits, demand a level of detail that is disproportionate to the fireworks budget — indeed the actual fireworks budget is often a relatively small fraction of the overall budget (and often the headline figure) for the event.

Such interested parties include:

- The event producers
- The creative director
- Sponsors
- Venue management
- Health & safety officials — internally and externally
- Fire experts
- Licensing authorities
- Other enforcing authorities (e.g., environmental officers)
- Landowners
- Media
- Security services

Any or all of these participants may, in our experience, have conflicting and contrasting demands!

The critical thing is to ensure that communication between all the interested parties starts early enough in the overall planning of the event so that the necessary elements may be incorporated early (perhaps even changes to the structure) and that there are no last-minute "show-stoppers"!

The role of the author's company, Davas Ltd, in the largest events has often been as an "honest broker" between the creative ideals and the practicalities of delivery. We ensure that what is required is feasible, and what is promised is safe. We also produce independent and objective assessments of the risks from the proposed display and the selected fireworks, and we provide independent and objective curtailment and cancellation criteria to be agreed on by all parties. In the run-up to and

Figure 20.1 – 10mm spider shell (Soldi) and 100mm white strobe shells (Zink) for Thamesday, London, 1989. Photo: Helen Saxton

during the event it is crucial that suitable contingencies be developed in case of adverse weather conditions or other factors affecting the viability of the display.

Each of the following events has required bespoke firing equipment to be designed and manufactured, then tested and proved before installation and the event itself. Very often either the fireworks have to be installed well prior to the event (for instance for the closing ceremony of the Olympic Games — prior to the *opening* ceremony) or in an extremely tight timescale immediately prior to the event.

Lastly, and sadly, the present threat of disruption of these national and international events cannot be dismissed. After all we are, by the nature of our job, placing explosives in relatively close proximity to large numbers of the public and smaller numbers of VIPs. Furthermore, we do our trade in the full spotlight of the world's media — so we just cannot afford to get it wrong!

The information provided by our various co-authors is obviously not a highly descriptive and commercially sensitive explanation of the event planning. It is meant to be a general overview of the challenges posed and solutions developed, which we hope will help those who stage such events in the future.

Planning the very large event

Very large-scale events using fireworks are many-faceted. Projects of this scale require complex stakeholder management. Public authorities not only approve and license the event; they are called upon to provide services and equipment. They must therefore be involved in the detailed planning and operations of the event. Networks of operations centres are set up to manage the event as a whole. The fireworks often become somewhat a minor part of the operation, which will include but not be limited to security, emergency services, policing, civil aviation, city-wide traffic management and street closures, public transport, inter-government agencies, tourism operators, suppliers to the project, the media, VIP site operations and cleaning and waste.

The creative team

The creative process for large-scale events with fireworks usually involves a great number of people across many disciplines. Typically this will include:

Creative director: Responsible for the idea and creative vision. The Creative Director will conceive, plan and/or devise the story of the show and work with the team as a whole to realise the vision. The CD works with the producer to realise the project within budget and operational constraints.

Music director: Often is also a composer who will select or compose music to underscore the fireworks. Selected music will be edited into a compilation and may include well-known tracks and anthems as well as "client" messages. There are, of course, classic fireworks tracks, perhaps commencing with George Frederick Handel's 1749 "Music for the Royal Fireworks". All sorts of classical movements and popular music are associated with fireworks shows. Original music requires composing and recording and may be prohibitively expensive in some cases. However, with the wide use of sampling and electronic composition and recording tools, original music is now within reach of many productions.

Fireworks designer: By combining knowledge of fireworks, music and the vision of the creative and music directors and taking into account site and audience constraints, the fireworks designer will select product suitable for the production and design effect sequences to create the show. The fireworks designer may also work with the manufacturer to create special effects and product specifically for a particular show.

Fireworks programmer: The programmer takes the product information and music track, and using computer technology applies the fireworks code to the tracks to inform the firing systems when particular shells or effects should be fired within the show. Working with the fireworks designer, he or she also designs the layout of the firing site and the electrical firing system.

Lighting and projections designer: Many fireworks or pyromusical shows also combine other effects, principally lighting and projections. The lighting designer will work with the rest of the team to incorporate lights and projection into the show. Often the fireworks track will also have a SMPTE timecode, which allows lights, projections and many other elements to be synchronized to the music.

Lead production team

Producer: Controls the project, runs the budget and is responsible for management of the project as a whole. The producer supervises compliance and permits as well as staffing and crew.

Operations manager: Arranges site access, and service, catering, health and safety, transport, equipment and PPE.

Safety officer/risk manager: Develops risk and safety management plans. Works with authorities and production personnel to ensure these plans are submitted and distributed to the appropriate authorities and personnel. Oversees risk and safety plans and management on-site before, during and after the event.

Stadium theatre events

Large-scale fireworks events can broadly speaking be broken up into two categories. Stadium-based events, "Stadium Theatre" and events staged in the public domain are fired in and around cities and landmarks. The use of architecture is often an integral part of these displays, mainly because the structures are part of a landscape or because the display is commissioned to "illuminate" the structure itself. Of course the use of stadium structures as part of "Stadium Theatre" almost goes without saying. The Patras Bridge and the Eiffel Tower are good examples of highlighted structures. The rooftops of whole cities are a great site for massed large-scale displays.

Large-scale citywide displays are often used as part of cultural or national celebrations or as part of a city or country's attempt to draw attention to itself, or indeed both. Sydney New Year's Eve is a good example.

Such displays provide for massed viewing from multiple vantage points. The different view and audience experience across the many viewing sights needs to be taken into consideration in the planning (this has been discussed elsewhere in this book) and if the show is to be synchronised to music, the distribution of the sound track to remote viewing sites also needs to be taken into consideration.

Stadium-based shows are in effect a truly theatrical experience. "Stadium Theatre" often uses fireworks in a very different way to that of the large-scale public-domain shows. The fireworks become part of the overall story. They provide punctuation to the narrative and often signal the end of a scene or chapter. They identify the page turns in the story and do not need to be extensive — often a single "hit" is sufficient.

"Stadium Theatre" also introduces additional complications, not least because fireworks fired within a stadium can often produce too much smoke, hence the future development of low-smoke devices.

If performers are to be moving mounting points for fireworks (and the range of such items is very restricted) then adequate precautions must be taken to ensure they are fully protected and properly trained, and that there are no undesirable interactions between performer and performer, or between performers and third parties. Again, such use is driving the need for new firework compositions — in particular those that are low temperature.

Event example – London Millennium

Description	London Millennium
Location	Central London — see map
Venue	Thames / Citywide
Live Audience	2 Million +
Broadcast Audience	3 Billion +
Production Team	200
Lead Agency	Ten Alps / Syd Howard Fireworks
Firework Designer	Syd Howard
Project Producer	Andrew Walsh
Production Manager	Tom Smith
Production Coordinator	Rosemary Walsh
Fireworks Programmer	Robert McDermott
Firework Crews	Approx 160 — see text
Music Director	Sir Trevor Horn
River Crew	Peter Duggan — General Marine

Creating a fireworks show along a 10Km section of a busy river in the centre of one of the world's great cities could be considered ambitious, and it was. However, in order to mark the turn of the millennium, that was exactly what London required. The show was to be part of a global broadcast and performed live to two million people along the Thames and broadcast around the world to an estimated three billion people. The sound track told the story of a millennium of English music down the ages. The story of the meridian and time were also critical to the event. The concept of the show was to fire the same show from multiple barges moored down the river. There was a large-scale chase effect, the "River of Fire", that ran the length of the display site, from Lambeth Bridge to Tower Bridge.

The Australian fireworks company Syd Howard Fireworks International was hired by the New Millennium Experience Company to produce the show. They provided the design, fireworks and the firing systems for the show. Working with Sir Trevor Horn, they also created the music track to which the show was cued.

The challenge was to find enough licensed and experienced operators available in the UK that could join the fireworks team. After some investigation it was decided to approach UK-based companies to provide racks and mortars and associated kit as well as operators. Eight companies were appointed to the project and joined the team to load and fire the show from individual barges assigned to each crew.

This proved to be an interesting experience. Eight companies, all extremely competitive with each other, many who had never spoken

Figure 20.2 – The Millennium team including Andrew Walsh, Trevor Horn, Robert McDermott and Tom Smith

to the others or were now not talking to each other, turned up at Centrepoint for a briefing. All these disparate groups, seeing a common adversary in an Australian fireworks company (for goodness' sake), formed an instant industry association on the spot. No longer were they opposing each other; they now had Syd Howard to take aim at and they did. The meeting was at times what can best be described as hostile. However, in the end eight UK companies joined the team and fired synchronised displays from barges as shown in Table 20.1 below.

Table 20.1 – London Millennium display companies and barge positions

	Place	Company	Comments
1B	Downstream of Tower Bridge	Happy Dragon	Moored
2A	Upstream of HMS Belfast		Held on tug
2B	Downstream of London Bridge	FTF Worldwide	Moored
3A	Upstream of London Bridge		Moored
3B	Downstream of Southwark Bridge	Fireking	Held on tug
4A	Upstream of Southwark Bridge		Held on tug
4B	Downstream of Blackfriars Bridge	Millennium	Moored
5A	Upstream of Blackfriars Bridge		Moored

	Place	Company	Comments
5B	Downstream of Waterloo Bridge	Pains	Moored
6A	Between Waterloo and Hungerford Bridges		Held on tug
7A	Upstream of Hungerford Bridge	Pyrovision	Moored
7B	Downstream of Westminster Bridge		Held on tug
8A	Upstream of Westminster Bridge	Merlin	Moored
8B	Downstream of Lambeth Bridge		Moored
9A	Upstream of Lambeth Bridge	Midland	Moored
9B	Downstream of Vauxhall Bridge		Moored

Finding a suitable secure location to store this large amount of fireworks was a challenge, as it needed to be secure and have a large enough exclusion zone to safely store the containers of product. A site in the then deserted docklands was chosen. The containers were stored on barges for added security, and for the convenience of moving them around within the docklands.

The decision was taken to keep the contents of the container secret and operations at the storage site low key, thus not drawing attention to what was being stored on the site, which proved an effective method of maintaining security. Loading operations took place at Royal Victoria docks prior to the 16 barges being moved into position on the Thames. This in itself was not without difficulty; King George V dock is immediately adjacent to London City Airport, but with good planning and proper risk assessment the case was made, and accepted, that the presence of the fireworks was no threat to the airport operations. A temporary exemption to the normal requirements for licensing the dock as an explosives factory was granted (although subsequently a permanent licence was granted), which detailed the way the fireworks were to be stored and handled.

Each barge was topped with a scaffolding and wooden platform. On Millennium Eve, the barges were moored in the river between bridges and in some cases held in place by tugs as shown in Table 20.1.

The Millennium project involved a very wide range of stakeholders: all the London boroughs involved, the City of London (GLA), the Port of London Authority (PLA), security agencies, the Health & Safety Executive (HSE), the London Fire Brigade (especially the river service) and the London Police. As it was a fireworks show of a scale never before staged in London (or anywhere else for that matter), many involved had little experience with the size and scale of the event. This required our production team to provide detailed briefings to explain the scope of

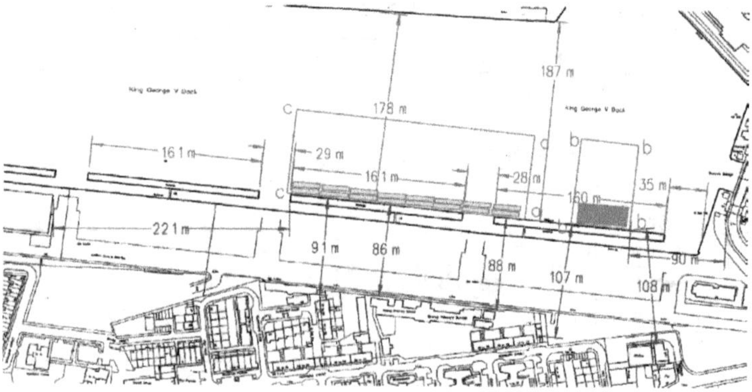

Figure 20.3 – Rigging area for barges in King George V dock — adjacent to London City Airport!

the project. The major challenges around the event were having people get to grips with the scale of the project and resourcing.

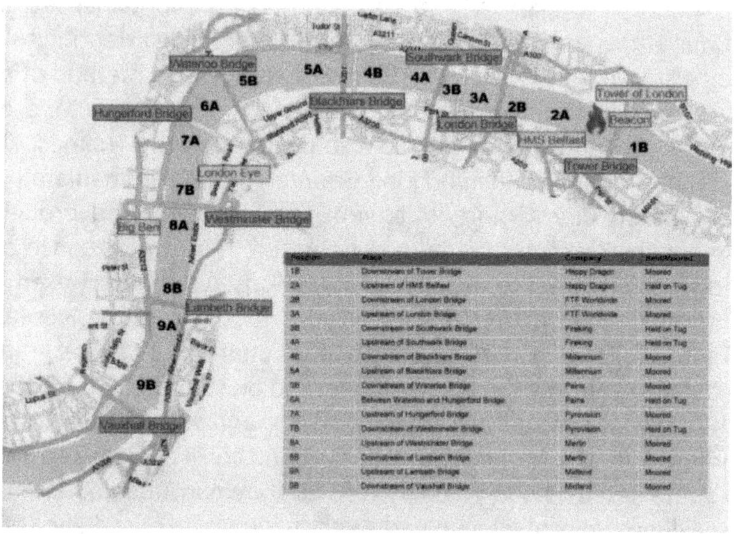

Figure 20.4 – Location of London Millennium barges

The project was fired on the stroke of midnight and was hailed as the best event of the new millennium.

It should be noted that as a result of badly planned media relations by an external source, some of the English press misunderstood the "River

Figure 20.5 – Barges loaded at King George V dock

of Fire" to some degree and at first lambasted the display. For a very small section of the press it would seem that the expectation around the first cue in the show was that the river would indeed burst into flame bank to bank. This of course was technically, practically and physically impossible (not least because of the danger of having flames licking at 16 barges containing tens of tonnes of explosives). Nevertheless some section of the media chose to portray the total event based on the expectation that the Thames would burst into flame. In fact, the concept for this section was a high-speed chase that ran down the river from Lambeth on its way to Greenwich at the speed of the rotation of the Earth, and the effect would have been breath-taking. Budgetary and practical considerations ruled this vision impossible to achieve. The lesson — make sure that what is being said to promote your show is realistic!

The practicalities of rigging 16 barges (and even of obtaining 16 suitable barges) for a display of this scale to be fired in the middle of an English winter cannot be over-estimated. The logistics of organising deliveries to each barge in turn, rigging and then securing each barge, and then ensuring that each barge was transported to its firing position with the minimum of wasted effort took many months of planning and a dedicated team of people — from the eight diverse firing companies as well as the organising group, barge men, river authorities etc.

Figure 20.6 – The "River of Fire." Note also the London Eye (not used for this display)

Event example – Athens Olympics 2004

Description	Opening and Closing Ceremony
Location	Athens
Venue	Athens Olympic Stadium
Live Audience	85,000
Broadcast Audience	4 Billion+
Production Team	600
Lead Agency	Jack Morton Worldwide
Creative Director	Dimitris Papaioannou
Executive Producer	Andrew Walsh
Fireworks Team	Groupe F
Fireworks Consultant	Tom Smith

Figure 20.7 – Athens Olympics 2004. Photo: Groupe F

Fireworks in the 2004 Athens ceremony were very much used as punctuation points and page turns that signalled new chapters in the flow of the show. They also enhanced dramatic moments within the show and built excitement. The first act of the Opening Ceremony was called "Calling to the Ancient Olympic Spirits", and the key effect was a comet that symbolized the fire of the ancients giving life to the modern Olympic movement, thus bridging the past and the present together. This was achieved by a line rocket fired from the stadium screen to the centre of the water-filled stadium, igniting the giant gas-fuelled flaming Olympic rings installed under the water. So a large-scale pyrotechnic effect opened the show.

Figure 20.8 – Location of Athens Olympics fireworks

The narrative of the ceremony was based around an allegory of traditional Greek culture and history harking back to its mythological beginnings, and viewed through the progression of Greek art punctuated by firework cues fired from the stadium roof. An enormous fireworks finale brought the show and the broadcast to a close. The fireworks very much used the structure of the stadium, two huge sweeping arches that reached over the entire structure. Mutual chase effects of comets and mines fired from the stadium arches were designed to animate the stadium while shells burst overhead.

Challenges

The challenges of using fireworks at an event such as the Olympics are many — not only are there many artistic demands placed upon the display designer by a wide variety of interested parties, but the event's biggest audience is not in the stadium but on worldwide television.

In addition, at Athens the logistical problems of rigging on the arches (and access to the arches) meant that specially designed brackets were developed that could be simply rigged, but incorporated failsafe elements so that once the pieces were in place additional rigging trips were not necessary even if a piece failed.

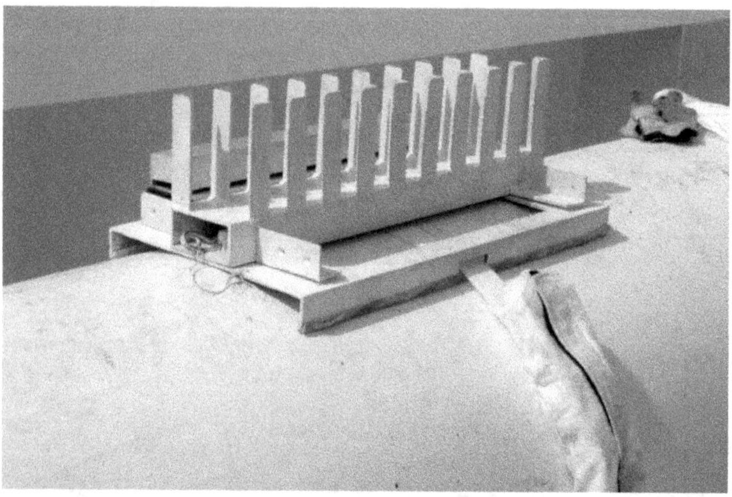

Figure 20.9 – Mock-up of fixing bracket for Athens Olympic Stadium arches

For firing aerial fireworks from sites in the "common domain" — the public area around the stadium through which the audience would pass on the way to their seats — it was decided that the mortars would only be positioned once the event had commenced so that no audience would be passing loaded mortars at what would be extremely short safety distances. To this end, the well-publicised problems of construction worked in our favour, to some extent — areas that were originally described as "white marble" became "compacted dirt" — much more suitable for erecting and launching a major display!

The majority of the fireworks very much used the structure of the stadium — two huge sweeping arches that reached over the entire structure and from the rim of the stadium roof — and were complemented by aerial material fired from the area outside the stadium itself. The majority of these fireworks were shot either vertically or angled slightly inwards.

In addition, there were some low-debris effects on the outside of the stadium roof, facing outwards, to give a symmetry and to extend the design for the exterior TV shots (they were not visible to the audience within the stadium).

Figure 20.10 – Athens Olympics 2004. Photo: Groupe F

Figure 20.11 – Athens Olympics 2004. Photo: Groupe F

The event was also notable for the non-pyrotechnic special effects within the stadium, including the flooding of the entire central area to allow a boat to move across the stadium, and the reproduction of the Olympic rings in fire (produced from gas burners as seen in Figure 20.10).

In many respects the Athens 2004 display is recognised as the fore-runner of the modern firework displays for ceremonies — the important elements being:

- Using fireworks for punctuation
- Extensive use of structures
- Extended chases on stadium and structures
- Less reliance on aerial components
- Compromise between the demands of the live and TV audience

Event example – Melbourne Commonwealth Games 2006

Andrew Walsh – Accolade Events
Tom Smith – Davas Ltd

Description	Opening and Closing Ceremony + River
Location	Melbourne
Venue	Melbourne Cricket Ground (MCG) & Yarra River
Live Audience	100,000 in MCG c. 500,000 along the Yarra
Broadcast Audience	3 Billion +
Production Team	550 (+ 10,000 volunteers)
Artistic Director and Executive Producer	Andrew Walsh
Music Director	Lyn Williams
Composers	Christopher Gordon Richard Mills Delta Goodrem
Fireworks Team	Howard & Sons
Firework Designer	Andrew Howard
Firework Programmer	Stuart Bensley
Fireworks Consultant	Tom Smith

For the Commonwealth Games in Melbourne we wanted to pay homage to the city itself, not by building some sort of representation of it in the stadium but by taking the whole show at times out of the stadium and into the public domain. One of the unique futures of Melbourne is that the MCG (Melbourne Cricket Ground) resides in a large parkland precinct adjacent to the central business and cultural districts of the city. Key to this realization was fireworks sites along the Yarra River and atop

Figure 20.12 – Location and layout of M2006 displays

of key city buildings that would illuminate the city and help tell the story of a sophisticated cosmopolitan city.

The fireworks along the river played the key element in the "Welcome to Melbourne" segment that showcased the city to the world.

Figure 20.13 – The display team working on the Melbourne Cricket Ground — M2006

Figure 20.14 – Pontoons on River Yarra with Melbourne city sky-line and buildings for part of aerial display in background

Fireworks were also used as punctuation points during the show to separate the arrival of the dignitaries and signal the start of the show. The official opening of the games also utilized large-scale cues from the roof of the stadium.

Figure 20.15 – M2006 display

Original music was commissioned for the whole ceremony and the composer was briefed to create scores that would provide cue points for fireworks.

In this way fireworks were used at many points throughout the whole event, but for relatively short durations — typically 5–15 seconds.

Figure 20.16 – M2006 display

To finish the evening, and once the VIPs had left the stage, a more typical five-minute display was fired, incorporating elements within the stadium as well as on the river and from strategic rooftops in the city, to create the most impressive extended display — see Figure 20.17.

Rigging the display/fireworks used

The majority of the fireworks to be viewed within the stadium were supported on the roof structure. The roof of the Melbourne Cricket Ground (MCG) is essentially made of strengthened glass panels (Figure 20.18) so that supporting plates were necessary to spread the recoil loads. Testing and calculations of such loads were done to assure the venue and the producers that the setup was safe.

The MCG is effectively circular, although the roof exists on two levels. This meant that whatever the wind direction, there was almost exactly the same likelihood of debris falling into the stadium. Once under the stadium rim it was very difficult to predict where it would fall. Calculations, discussed in Chapter 7, showed that minor debris would fall inside the stadium, but that there was no possibility of lit debris falling onto

Figure 20.17 – M2006 from the air. Photo: Howard and Sons

Figure 20.18 – Melbourne Cricket Ground — rooftop working

the audience, or for projected effects from a malfunction reaching any person within the stadium. Furthermore, the fireworks were chosen to produce minimum debris, and that debris was examined to determine the potential for harm should it fall in the audience area. As a result of the work done and the calculations performed, the display was allowed to take place with the acceptance that some minor debris would fall in the areas containing performers, athletes and the audience.

For the final scene in the ceremony we wanted to create a mini-show within the show for the athletes. Since the athletes are all about speed and endurance, we made a high-energy segment featuring a song commissioned from Delta Goodrem. The segment featured high-speed blade skaters in a pyrotechnic ballet. Each skater carried a pyro-backpack, and these specially designed backpacks held several fireworks cues that were fired by the skaters according to specific cues.

Figure 20.19 – M2006 — Rollerbladers with gerbs and saxons attached

Each pack had three elements fired on separate cues:

- A single gerb
- Changing to a fan of gerbs
- Changing to a saxon

The final effect can be seen in Figure 20.19.

This was one of the most complex elements to design and to demonstrate low risk for. The design and safety of the skaters' backpacks included a safety plug that prevented the fireworks from being fired inside the stadium building prior to entry into the central arena. This plug was only handed out to each skater by the pyrotechnicians as the skaters entered the arena.

Event example – London New Year's Eve

Jim Donald – Jack Morton Worldwide
Tom Smith – Davas Ltd

Event	London New Year's Eve
Location	London
Venue	London Eye and barges/pontoons on the Thames
Live Audience	250,000
Broadcast Audience	100 Million +
Production Team	200
Project Direction and Management	Jack Morton Worldwide
Producers	Robert Alge & Jim Donald
Firework Company 2004–2009	Groupe F
Firework Designer 2004 – 2009	Christophe Berteneau & Jonas Bidault
Firework Company 2010	Kimbolton Fireworks
Firework Designer 2010	Darryl Fleming
Lighting Design	Durham Merenghi
Pyrotechnic Consultant	Tom Smith

This display, now known the world over, is centred on the structure of the London Eye, but in order to add spectacle and variety, and to minimise the amount of time spent and pyrotechnic content on the Eye itself, a number of barges and other floating craft have usually been employed on the river Thames in front of the Eye. The use of such barges has changed each year and has included the simultaneous use of a large number of small firing platforms on purpose-built catamarans connected to, and held between, the main firing barges, and the use of specially adapted "pyro boats" moving up and down river during the display. In both cases these platforms fired only relatively low-calibre and low-debris effects such as Roman candles with comet stars.

The London Eye itself presents significant challenges as a firing structure — not least because it is a wheel, and rotates (although for firing it is "locked down" in a single position).

The London Eye is a Ferris-wheel structure on the south bank of the Thames in London, more or less opposite the Houses of Parliament. It was unique in construction in that the gondolas or capsules (never "pods") are on the outside of the wheel structure — affording the persons inside the best view, but causing some problems when designing a firework display!

Between each capsule there is an extensive lattice-work support structure and this, together with the rim of the wheel, is the firing point of choice for both outward-firing and inward-firing effects.

Figure 20.20 – Layout and location of London New Year's Eve displays 2003–2010

Obviously, none of the effects must impinge on the structure when firing, and great care must also be taken to ensure that the rigging methods do not damage or mark the structure itself.

In addition to the positions on the outside of the Eye the central hub has also been used for pyrotechnic effects.

Figure 20.21 – The London Eye

Types of fireworks used on the London Eye

In addition to the normal aerial content fired from the Thames a specialist range of material has been chosen to fire from the structure itself. In all cases the fireworks have been selected for the following features:

- Low debris
- Low dross effects (i.e., clean burn with no materials to impact on the structure of the Eye or on the material of the gondolas)
- Precision of firing and of performance (so that the patterns produced are accurate)

The types of fireworks used include:

- 18, 25 and 30mm Roman candle single shots
- 30mm and 45mm single-shot mines
- 19mm silver jets (short-duration gerbs)

Outline

This annual display, fired on behalf of the Greater London Authority (GLA) and the Mayor of London, centres around the London Eye.

Figure 20.22 – The hub of the London Eye

Figure 20.23 – London Eye lattice structure between "capsules"

The current display, a project managed by Jack Morton Worldwide, has run in 2004/05, 2005/06, 2006/07, 2007/08, 2008/09 and 2009/10. Prior to that there were displays on the same site for the Millennium celebrations and in 2003/4, which were organised by other companies. The author has been involved in the planning and execution of each of these displays.

In addition to items fired from the London Eye structure itself, the display is complemented by aerial fireworks fired from barges on the Thames.

The main audience for the display is, in fact, the millions of people watching the display throughout the world — not only because of the display itself, but because of the geographical significance of London as the centre of world time. In addition of course, the event attracts a very large number of members of the public on the banks of the Thames and some of the bridges adjacent to the display site. The market for this event is very broad, touching members of the public of all ages and backgrounds.

Background

In 2004, the GLA wanted to create a New Year's Eve experience that would surpass any other, to give something back to Londoners and show the world that London would be the perfect host city for the 2012 Olympic Games. In just five years, the Mayor of London's New Year's Eve celebrations have become famous worldwide as awe-inspiring experiences, on a par with long-established events in Sydney and New York. Each year, our brief is to use the New Year celebrations as an opportunity to promote London as a world-class city, raising its profile as a national and international tourist destination. Each year sees a new and relevant theme being explored in our production.

In 2008/09 the GLA secured a sponsor for the first time, which means that in years to come the creative design will be adapted around the sponsor's messaging. This will enable the event to remain financially sustainable. In 2009/10 the contractor was selected by competitive tender for the first time — which resulted in a change of supplier.

The thinking and the experience

Each year we have set out to create the best possible spectacle — dramatically blending fireworks with projection and lighting. A powerful "added-value" innovation has been the integration of a narrative thread into our performances — a simple thematic underscore — and the subject matter changes year to year.

Each year we have worked closely with the GLA to develop an experience that would invoke pride and delight in Londoners and create a "media moment" that would capture the front pages of newspapers worldwide.

Each year sees a new and relevant theme being explored. This concept is brought to life in a stunning ten-minute firework display, supported by a full multimedia experience of large-scale projections onto the Shell building, an audio soundscape and a spectacular lighting show on and around the London Eye. This amazing and instantly recognisable landmark in the city is more than just a backdrop for our festivities — it is the canvas for our communication.

The impact

In 2004, some 150,000 attended the event. In 2008 we had over 250,000 people brought together on the banks of the Thames and surrounding areas, with upwards of 10 million national TV viewers, showing how the success of the experiences over the years has attracted more people, including worldwide press/online media coverage.

Figure 20.24 – Comets and silver jets. Photo: Jack Morton

The practicalities

With the increasing number of people coming to see the event, the planning focus has changed. Creating this world-class spectacle now includes the intricate detail of developing a full crowd-management plan for a large section of Central London, ensuring that viewing areas are filled to a comfortable capacity, and that once those areas are full, members of the public are directed to parts of Central London where there is still space to view safely.

In addition, we manage a comprehensive community liaison and permissions plan to ensure that all key stakeholders, whether local residents or businesses, have a full understanding as to the impact the event and increased crowds will have on their lives. This process takes months of planning and liaison with the multiple agencies that run London on a day-to-day basis. This group includes the local authorities, emergency services, transport networks, local business groups and landowners. This liaison work is carried out to ensure that all agency plans are aligned and interlinked throughout Central London and to maximise the enjoyment of both visitors and local residents.

In addition to the crowds, our remit is to minimise the impact of the event on London, and this extends to both the show itself and the crowd-management challenges. One of the problems is that the fireworks and lighting teams have huge restrictions on when they can work to rig the show. The London Eye needs to be open to the public every day through to 4pm on December 31st, and no live fireworks can

Figure 20.25 – Shells and mines from barges and pontoons. Photo: Jack Morton

be rigged onto the Eye before this. Furthermore, any pre-rigging to attach the brackets and wiring has to happen during the night from December 26th through December 30th. Adverse weather, which is highly probable at that time of year, can hugely affect the riggers' process.

Then, on December 31st at 4pm, both the lighting and the fireworks teams have to work seamlessly together to load moving lights into each capsule and load the live pyrotechnics onto the pre-fixed bracketing. This process has been honed down over the years and is now a very slick operation, allowing for testing and lockdown of the Eye to be completed by 10pm, just two hours before the display is fired.

Rigging the display

The practicalities of working on a structure such as the London Eye are significant, not least because every minute the Eye is out of commission costs a significant loss of revenue.

In order to minimise the time spent rigging on December 31st, the fireworks are now mounted on removable brackets, while the part of the bracket fitted to the structure of the Eye is installed during the nights leading up to the 31st. In 2003 we adopted a different approach in which custom-designed "pyro boxes" were installed in the run-up to the 31st, sealed (they were tested to withstand accidental ignition) and only opened on the 31st. Both approaches have their merits, but both pose significant technical challenges. The decision about which route to take depends on many factors:

- The type of structure
- The time that may be allocated to rigging
- The location and time of year — and the likely weather conditions
- The weights of the brackets or firing containers
- The durability of the brackets or firing containers
- Possible failure modes, impacting on the structure, the riggers and the public — during the rigging and de-rigging phases

On other structures the approach taken may be different from those developed for the London Eye.

Wiring the London Eye display usually takes place prior to the installation of any live fireworks, but again on other structures this may be different. The important factors to consider in rigging a display on a structure include:

- The fact that very large cable runs may be needed — which require supporting and the possibility of induced current from the large cable runs
- Possible interference from RF and other sources built into or fixed onto the structure
- Ease of making the final fixing from the firing lines to the installed fireworks — which may have to be done in a very short timeframe, as they are on the London Eye

Figure 20.26 – Gold comets fired from three barges. Photo: Jack Morton

It is tempting to consider the use of wireless systems to overcome some of these concerns, but care must be taken to ensure that they are physically robust enough for the job and that they are not susceptible to interference causing misfires or disrupting the firing sequence. Our rule of thumb — "If it won't work in the middle of a wet field in November in the UK, then it's not worth using" — is a reasonable starting point in the decision-making process!

Figure 20.27 – Mines and shells from barges, comets from London Eye. Photo: Jack Morton

Figure 20.28 – Shells from barges. Photo: Jack Morton

Figure 20.29 – London New Year's Eve. Photo: Groupe F

Awards

The award-winning event has been honoured most recently with the "Creative Partnership Tribute" from the National Outdoor Events Association (2008) and has also won "Public Event of the Year" at the Eventia UK Awards (2007) and the Event Gold Award for "Best Celebration Event" (2005).

Chapter 21 – Future developments

Tom Smith – Davas Ltd

The future of firework displays worldwide, and of future developments in delivering those displays, is as exciting and challenging now as at any time in the past. Despite the anti-firework lobby, opportunistic as they are — switching their concerns from safety to noise to environment — there is a huge public demand for fireworks at all scales of events.

The concerns should not be dismissed completely, though — they prompt the industry to continually innovate and improve in order to address those concerns, as well as to produce more interesting displays with novel effects.

There are significant challenges ahead, and although this chapter can only briefly address some of them we all know that as soon as one challenge is resolved another will take its place.

Supply

At present a large proportion of fireworks are manufactured in the Far East and exported all around the world. To a greater or lesser extent this switch from localised manufacture has been driven by price, and the major factor in the price is labour. Far East labour is, at present, incredibly cheap and the manufacture of fireworks is incredibly labour intensive.

Domestic manufacturers (and here we are primarily considering Europe, the US and Australasia) have progressively switched manufacture of simple, standard items to the Far East but have concentrated on manufacturing a range of specialist items that share some of the following characteristics:

- Higher precision — especially in the timing of multi-shot Roman candles
- Better colours, with a greater variety
- Lower smoke generation — primarily for aesthetic reasons, but increasingly for environmental reasons
- Single-shot, high-precision devices with strictly defined performance characteristics (e.g., all stars in a mine extinguished at a precise height) — especially for indoor or on-stage use, or for firing from structures
- Production of items difficult (or expensive) to transport from the Far East (often because of restrictions on transport of high-hazard items)
- Automated manufacturing techniques — to improve consistency and improve worker safety

In some cases it has been possible for domestic manufacturers to adapt sufficiently to maintain their workforce, and by so doing maintain the expertise developed over many years. In other cases this expertise has been lost forever. If, in the future, export of items from the Far East is further reduced in terms of devices actually produced, restrictions on transport or the economic demands of local workforces, the advantages that now exist will disappear and the products will be less attractive financially.

European or other legislation may actually exacerbate the problems and potentially lead to an increase in accidents and the inevitable resulting pressure to introduce even further controls. For instance, the European Directive on Pyrotechnic Articles (EC/2007/23) does not require persons importing fireworks for their own use (i.e., not placed on "the market") to carry out the same expensive and time-consuming product tests that they would have to if the items were sold commercially. The fear of many is that this approach (which is understandable, if perhaps misguided) will lead to a flood of smaller importers with less experience and training, importing lower-quality product, which will lead to an increase in incidents, which will lead to further controls on ALL. There are important principles here — a professional display company designing, rigging and firing a display IS NOT selling fireworks *per se*, they are selling a service.

On the other hand, calls by some to restrict the ability to import except to a few larger companies contravenes basic "free market" principles and effectively presents a barrier to trade.

Manufacture vs manipulation

Many companies who style themselves as "manufacturers" are really "manipulators" — that is, they take existing firework articles and rig them and fuse them in such a way to "add value" and to produce a new article. There is nothing inherently wrong in doing this, but companies must ensure that by creating a new article they take on the responsibility for its classification, transport issues, and the responsibility for compliance with other regulations if they apply.

Restrictions on supply, financial and practical, will lead to an increase in this type of activity, and care must be taken by regulators to ensure that their well-intentioned proposals do not adversely bias or affect the market in a way that leads to an increase in incidents and accidents, otherwise there is a danger of creating a self-fulfilling prophecy: fireworks are dangerous — so we need more regulation — which makes illegal (or semi-legal) activities more likely — which leads to more accidents!

Chemistry developments

The chemistry of fireworks has remained relatively unchanged for many years, and the fundamental designs of firework articles have changed even less (CT Brock would certainly recognise all the firework types available now). The main developments in displays in the last few years have been in increasingly complex and precise firing methods.

However, this is set to change in the next few years. We believe that there will be a drive to cleaner-burning, lower-smoke compositions that will have the dual benefit of appearing more attractive and reducing environmental pollution. Not that the environmental card should be over-played — the amount of gaseous and solid products produced from the combustion of pyrotechnic compositions within firework articles is very low in comparison with many other industries, and also in comparison with the combustion products from the internal combustion engines used to transport people to a display!

For the London New Year's Eve display, for example, it has been calculated that fireworks contributed less than 0.3 percent of the total environmental impact of the display — the remainder being almost all in transporting the audience to the display itself. Of course there will be the naysayers who would argue that if the display didn't take place at all the remaining 99.7 percent would not have occurred either, but then perhaps their objections can be dismissed as being as tenuous as their grip on reality!

Table 21.1 illustrates the likely chemistry developments in the next few years.

Technical developments

In addition, there are some technical developments we can foresee. In many cases the new technology makes something that was once technically difficult (but not impossible) available to a much wider set of users — but in some cases allows a completely new approach to be adopted. We believe that, like the development of pyrotechnic chemistry outlined above, much will be driven by a need to reduce risks, be they to human health or to the environment. The most exciting developments, however, are likely to be in the precision of the use of fireworks and the consequent possibilities for the display designer.

Table 21.1 – Chemical developments

Technology	Comments
Nitrocellulose-based compositions	Particularly "cold-burning" compositions that reduce the risk of fire or injury in close-proximity effects. However, they may find wider applicability in fireworks too.
High nitrogen–based compositions	Some work on "Green Pyrotechnics" in which the need for a separate oxidiser/fuel/colour agent combination is removed has already been carried out [67], although it is unlikely that such compositions will be commercially available or widely adopted for a considerable period.
Other low-smoke compositions	This is an exciting and interesting area — particularly for the display designer firing from structures where "unintentional smoke" can be a real problem. We foresee that developments in this area will come, in part, from the demands of broadcast media where smoky effects are all too apparent and distracting rather than from the live audience for whom some smoke (and smell) can actually enhance the overall effect!
Nano-compositions	Rather inevitably we expect the "nano" moniker to start appearing on fireworks in the near future. Whether this portrays a real push to nano-technology (which in the explosives field creates as many problems as it potentially solves — nano-particles are often pyrophoric!) or is just clever marketing remains to be seen. Potential developments that utilise nano-chemistry and rely on atmospheric oxygen for oxidation could produce some very interesting novel effects — but again these are unlikely to find wide adoption for some time.

Table 21.2 – Technical developments

Technology	Comments
"Air launch" and other systems	Disney developed an "air launch" system some years ago in which the blackpowder lifting charge of shells was replaced with compressed gas. Although this has not been adopted widely there is merit in using less polluting means of projection of, particularly, shells.
Precision effects	There is an increasing development of "fast-burn" effects where the pyrotechnic star or effect burns to completion before the visual effect "curves" too much from gravity. This has been particularly used in single-shot comets and mines to produce a linear column of stars or effects.
Precision placement in the sky	By varying launch time, launch angles and delays it is possible to produce precise 3D placement of effects in the sky. Although this work is relatively embryonic, the effects can be very dramatic. We look forward to further developments and wide-scale adoption in due course.

Technology	Comments
Precision timing	The development of precision timed bursts (as outlined in Chapter 18) also permits the designer to use fireworks creatively in a way that has not been possible before.
Firing systems	As outlined in Chapter 18 there are promising new developments in firing systems. There are some concerns about the use of wireless technology — for example, the possibility of accidental triggering by radio interference or other sources, or a system being used where the radio-frequency spectrum is restricted. Care should be taken to ensure that any proposed system is legal in the intended place of use and is "failsafe" as far as possible.

Legal developments

We also foresee some legal developments in the next few years that will affect the way fireworks are manufactured, used, stored, transported or used.

Table 21.3 – Legal developments

Technology	Comments
Flash powder	We foresee changes and refinement to the existing pressure/time (p/t) tests and hopefully the development of a list of permitted compositions (perhaps those that are definitely "flash", borderline cases that merit further investigation, and those that are definitely NOT "flash") that will allow manufacturers to work without having to test every possible composition.
Manufacture regulations	We perceive that regulatory bodies, especially in Europe, would like to harmonise the requirements for explosives manufacture — although we believe that trying to do so in anything but the most general way is fraught with difficulty and would achieve little. There is a great danger of a "lowest common denominator" approach that will suit no one, inconvenience most and critically achieve little in improvement of safety.
Transport regulations	One day we can actually see that the UN recommendations will be applied equally and equitably and that mutual recognition will indeed produce a transport regime that provides open borders — but we are not there yet!
Storage regulations	Like manufacture, there are increasing demands to harmonise provisions. Similarly there are potential problems.

Technology	Comments
Use regulations	Custom and practice vary so widely, it is almost inconceivable that harmonised regulations could be formulated and be meaningful except in the most general terms.
Security Regulations	There are proposals to increase the security provisions for all aspects of explosives, and obviously this is one area where harmonisation makes sense. However, we do feel that the security implications from fireworks are low — fireworks and firework components are simply not that attractive to terrorists.

Chapter 22 – What happens when it goes wrong?

Tom Smith - Davas Ltd
Murray Torrible - Precision Broking

We all hope that we will never have an accident at a firework display, but the old adage "if you work long enough with explosives you WILL have an accident" almost certainly applies. It is our sincere hope that this book goes some way to help you in reducing the likelihood that you will suffer that accident.

Accidents range from minor to catastrophic and can affect firers, the audience, innocent bystanders and even structures.

Accidents and incidents

Any accident or incident is bad news. It can obviously affect those directly concerned, affect the reputation of the display company and can often lead to expensive and time-consuming investigation and prosecution. The reputation of the venue and any sponsors can be similarly affected. Any firework accident also increases the demands to more highly regulate the industry (or even ban fireworks completely). In most cases the law is already perfectly adequate — applying the law proportionately and to the right people is the hard part!

It is to be hoped that the principles outlined in this book will lead to fewer accidents and incidents — but we are not naïve or arrogant enough to suppose that by adopting these principles all accidents and incidents will be eliminated!

Accidents involving fireworks will continue to occur, whether they be at the point of manufacture, during manipulation, transport, storage, rigging or use. They may occur to the person manufacturing, to the person firing, to their intended audience (or to structures) or to "innocent bystanders".

When any accident or incident occurs it is important to know how to react and what to do to minimise its consequences, and to gain information to minimise the possibility of something similar happening again (to the same people, or within the same company, or within the industry as a whole). The firework industry is relatively small, and an accident that affects one part of it often has consequences for the entire industry.

First aid for firework incidents

This book is not intended in any way to be a first aid manual — there are many fine texts available and many courses given on the basic principles of life-saving.[1] In general, the modern principles of first aid are simple — they only require the attending party to preserve life where it is threatened before more specialist advice (in the form of paramedics or doctors) can take over.

However, there are aspects of any firework incident affecting persons that deserve special attention. This is because pyrotechnic injuries usually:

- Are as a result of impact of a firework or sub-component, often at high velocity and sometimes with significant kinetic energy
- Can introduce chemicals into the wound, either from the composition itself or from combustion by-products
- Are often extremely hot — firework stars burn at temperatures up to 2500°C

Of these the last two may be regarded as pyrotechnic-specific, and it is probably only the last one that could directly threaten human life before specialist care is available and for which measures can be taken by the first-aider. Trauma from impact should be treated like any other impact, and in general the chemical issues will not be life threatening (although they should be raised as a concern to the paramedics and doctors).

In the case of any burn the most important action that can be taken is to cool the burn to lessen the possibility of damage to surrounding tissue. It is also important to realise that the damage can continue long after the patient thinks they are now "fine" and have been cooled sufficiently. This is why the standard advice is to place the affected area under cold running water for at least 15 minutes. If the burned area is large, be careful not to cause hypothermia by cooling the casualty excessively. Keep the unaffected areas wrapped in a coat or blanket. Remove jewellery and clothing except for materials that are stuck to the burned area. Leave these in place for qualified staff to tackle later. If blisters have formed, leave them intact. Cover the affected area with something that can be removed later without sticking. Cling film is very suitable, or a clear plastic bag. Be careful to avoid wrapping cling film too tightly around a limb; apply the cling film lengthways up and down the limb and then push it gently into contact with the skin or flesh. A clear plastic bag can be used to cover a hand or foot. A burn that is covered in this way is much less painful than one that is left open to the air. If a burn is more than a few centimeters in area, has caused blisters or deeper tissue damage, is on the face, hands, feet or over a joint — or if the person has a complex medical background, or is a child or elderly person — always seek medical assistance.

Information on-site

If any incident occurs it is important to gather information on the display site that could be of subsequent use in an investigation (either informal or formal) or to defend a claim, or from which to learn lessons.

The basic information should include:

- Where the incident occurred
- When the incident occurred (both the time and the point during the display)
- Who was affected (names) and where they were located
- A site plan indicating the firing area, the incident area, distances and the wind strength and direction
- A site plan showing the layout of the firing area and indicating the firework that caused (or is assumed to have caused) the incident
- Photographs of the display site (generally cell phone pictures are sufficient for preliminary information)
- List and contact details of firers, assistants, organisers, stewards and any witnesses

Fireworks from an insurer's viewpoint

The first topic to discuss is liability and insurance. Without adequate insurance, you or your company will be potentially liable for the costs of legal defence, third-party costs and damages if they are awarded. It is important to note that claims are expensive whether you are liable or not.

Anyone with experience of dealing with solicitors knows it is quite easy for costs to escalate in a few meetings or following a few letters. Claims that appear at the outset to be relatively minor can escalate with prolonged legal argument and expert opinion to become expensive claims with relatively small settlements and very large costs. It is also true to say that the most careful pyrotechnician cannot guard against all eventualities or the public's ability to claim for perceived injury. Over the past ten years the UK firework industry has seen claims from third parties arising from:

- Rampaging cows
- Non-hatching of pheasant chicks
- The loss of income caused by firework-induced post-traumatic stress of a potentially race-winning greyhound
- Injury sustained following insertion of a firework into an orifice in which a firework should not be inserted

Definitions

Liability means legal responsibility for one's acts or omissions. It is derived from English tort law, tort meaning injury, which is the basis for civil legal actions in the UK. The police deal with matters that are prosecuted by the state, but liability mostly deals with injury between individuals, which are civil matters rather than criminal. Liability cases that go to trial are heard by a judge but not in front of a jury.

The law perceives companies and individuals as having "a duty of care" to people who would be affected by their actions. The duty of care varies from person to person depending on circumstances and the person's ability to protect themselves or to choose to avoid a situation. As an illustration of the variation, one's duty of care is greatest with infants and reduces with adults. It is greater with employees and less with members of the public.

Operating a firework company gives rise to a number of legal responsibilities and creates a duty of care to anyone that could be affected by the company's operations whether the duty of care is owed to an employee working at a storage facility or a member of the public on a contract site.

Insurance

It is useful to explain why a firework company would buy insurance. Many companies view insurance as another tax, not fulfilling a practical function but necessary in order to comply with regulations for storage or for tender requirements. However, the real reason that firework companies should have insurance is to protect them from the huge costs of litigation. You may have operated your firework company for many years without claim, which may be skill or good fortune. However, if something does go wrong and someone does take action the costs are considerable. During the past ten years the industry has seen soft-tissue injuries escalating into settlements in the region of £80,000 or what seemed to be harmless grass fires escalating into settlements in excess of £250,000. Liability insurance offers protection whether or not the action taken by the third party is successful, picking up the legal defence costs and the loss adjuster's costs of investigating the claim and mediating with the third party. Unfortunately, within the UK it is not unusual for individuals to sue for an actual or perceived injury whether we would consider it justified or not.

UK insurance market

The UK is similar to the rest of the EE, the US and Australia in that there are a limited number of insurers willing to insure the firework industry.

The reasons for this restricted market is due, in part, to the perception by the insurance industry that working with fireworks is high hazard. The second and probably more influential reason is that the firework industry is relatively small and the insurance industry makes more money in other sectors.

In the UK there are two online insurance markets for one-off events and predominately three insurers providing annual liability policies for firework companies. Each of the three insurers tends to deal with one broker to distribute their insurance policies, resulting in firework companies selecting an insurance product from one of three insurance brokers.

What cover do you need?

There are three basic types of insurance cover:

Employer's liability

If your company employs people other than yourself and immediate family you are required to have Employers' Liability insurance by the Employers' Liability Act 1969 in the UK. If you operate a firework company it is almost inevitable that you will require Employers' Liability insurance. The courts have now interpreted the term *employee* so widely that it extends to mean "people who are borrowed or loaned," so simply asking someone to lend a hand moving something, by current legal definitions, makes you an employer of that person or owing a duty of care as an employer to that person.

The insurance market offers employers liability with a £10 million limit of indemnity. A limit of indemnity is the maximum amount that can be claimed per incident. This will either be expressed as "per occurrence", which means that there is no annual limit on how many incidents in any one year, or it is expressed as "per occurrence and in the aggregate", which means it is the maximum that can be claimed for any one incident and during the policy year. The policy cover is broad and liability is described as strict. The meaning of *strict* in this context is that because an employee is taking instructions from their employer, the law assumes that the employer is providing a safe work environment, training and tools. If there is an incident, by definition the employer has failed in his duty of care and is liable.

Cover under Employers' Liability policies is broad, extending cover to include driving to and from work and working on temporary contracts overseas. There is rarely any policy excess under Employers' Liability policies.

Public liability

Public liability insures you against actions taken by third parties for damage to their property or injury to their person allegedly caused by your negligent actions or omissions. Unlike employers' liability, public liability is not a legal requirement in the UK. It is legal to operate without public liability, however, so in practical terms it is not possible to operate without a policy, as HSE storage clients will require cover to be in place.

Most public liability policies in the UK offer cover that is limited to any one occurrence but unlimited during the policy period. This means that the insured can have multiple claims during a year and the policy will continue to operate. This differs from something like a motor policy where if the car is a write-off you effectively have to arrange a new policy, as the insurance contract has been satisfied.

The UK insurance market offers policies ranging from £1 million limit of indemnity to £10 million limit of indemnity. Most firework companies in the UK have a £5 million limit of indemnity as a result of insurance market preference rather than as a reaction to claims made against the industry. Public liability claims are usually subject to an excess, which is the amount of liability retained by the insured. Excesses vary from insurer to insurer but usually start in the region of £250 and can be as high as £5000.

Product liability

Product liability insurance offers protection for claims made by third parties following damage to their property or injury to their person caused by the operation of a product. Within the firework industry the greatest exposure to claims is to manufacturers and importers. Importers in the UK are treated as manufacturers, as it has not been possible to succeed in a product liability claim against Chinese manufacturers. Liability is relatively strict with products because people buying the product only have the information supplied with the product to inform them how the product should be used.

Product liability policies' limits of indemnity tend to match the public liability limits. However, the limits are per occurrence and in the aggregate, which means the maximum that can be claimed during any one policy period is the stated limit of indemnity. Product liability claims are usually subject to an excess.

Operation of a claim

Let's say the worst has happened. Something has gone wrong and there has been an injury at a display; a firer or member of the public has been injured. What do you do? Unless the incident is minor the display needs to be halted. The priority is obviously ensuring that the injured person receives the correct level of medical attention as soon as possible. This might mean making sure they see the St John's Ambulance on-site or it might mean that stewards ensure access by emergency vehicles to the accident site.

Do not admit liability to anyone. This is not a cynical avoidance of blame, but where there is an injured person you feel sympathy and want to make the injured person feel better. Unfortunately, in our society where you apologise it is seen as an admission of fault and it will make the insurers' job more difficult when they are trying to establish the facts in order to defend an allegation of negligence on your part.

Once the injured person has been dealt with, the accident site needs to be preserved and recorded.

The area where the person was injured should be photographed before the site is cleared or any load out commences. The firing site needs to be photographed. If it is possible at this stage to identify which effect caused the injury, the spent firework and supporting structure need to be photographed.

You should interview witnesses and crew immediately after the incident and record in writing their statements regarding the circumstances of the incident. Take down the contact details of all people who witnessed the incident in case you need to speak with them again or provide their information to loss adjusters or solicitors.

Where the incident is very serious it is likely the site will not be cleared until investigating authorities authorise it. If the incident is less serious and the site can be cleared, it is important to preserve as much as possible the product and supporting structure when transporting it so that it can be examined by loss adjusters and consultants at your premises.

Notification of a claim

Depending on how serious the claim is, notify your broker — either immediately or as soon as possible.

The broker will notify the insurer and the insurer will appoint an adjuster and — subject to the seriousness of the claim — a solicitor.

The various roles in a claim are:

Broker – Your broker's role is to monitor the claim progress and ensure that your interests are being defended. This can mean simply making sure the claim progresses smoothly or arguing interpretation of policy cover with adjusters and insurers. There are grey areas in insurance wordings, which a broker needs to ensure are interpreted in favour of the client rather than the insurer.

Loss adjuster – The loss adjuster is appointed by the insurer, and their role is to protect the interest of the client and insurer and to negotiate resolution of claims in the most effective manner.

Solicitor and/or barrister – Solicitors are appointed on serious claims and employers' liability claims or where litigation escalates and requires court representation. If the level of court is higher, barristers will be instructed.

Consultant or expert – Consultants are instructed where specialist services are required. This could involve testing of material or private investigation.

Insurer – The insurer stands in the position of the firework operator to defend the claim. They instruct the adjusters and solicitors and ultimately either successfully defend or settle the claim.

Third party/third party solicitor – The injured person will either contact you directly to claim that you were negligent and caused their injury, or they will instruct solicitors to represent them and to claim against you. They will allege that you were negligent and that you should compensate them or their client for their injury, medical expenses and loss of wages if they are prevented from working.

Claim investigation and the role of the adjuster/solicitor

The loss adjuster's role is to establish the facts of the incident and establish the incident's causes.

Once these points are established the loss adjuster will advise the insurer whether to settle a claim that arises out of incident. Where it is clear that the firework operator was at fault or the product was at fault the loss adjuster will advise the insurer to settle in order to avoid increasing the loss with unnecessary legal expenses. In these circumstances the adjuster will negotiate with the third party or their representatives and agree on a financial settlement. Where the circumstances are unclear or there is no evidence of fault on the part of the firework operator, the loss adjuster will gather further evidence to defend against claims in order to deny liability or to argue a smaller level of fault on the part of the firework operator.

Solicitors become involved in claims where the injury is serious or where the claim is being defended in court. Subject to the level of court it is possible that barristers will also be appointed.

Claim investigation

The loss adjuster not only has to investigate the circumstances surrounding the incident, but also in order to defend against a claim that the firework operator was negligent and caused the incident he/she must establish that the firework operator had safe working practices. In order to establish safe working practices it is necessary to have documentation that sets these practices out. If a company works very safely with excellent working practices but these practices are not documented, the third party's solicitor or the court will assume that they do not exist. A finding of negligence could be established where it was not warranted.

The loss adjuster will be looking for:

- Health and safety policy
- Risk assessment/method statement
- Spectator distances
- Fallout zones
- Operator guidelines
- PPE — including sign-off for equipment
- Pre-display report/survey
- Product list/manifest
- Post-display report
- Personnel records
- Training logs

As an illustration of the value of documentation: in the circumstances where an employee took off his head protection during a display because he did not like wearing it and subsequently was injured and suffered a burn to the side of his face, it would be important to show that the company required their staff to wear personal protective equipment. It would not be possible to prove this fact to a court unless the company's documentation stated that firers were required to wear PPE during displays, and ideally, that the equipment for that display was issued to the firer and signed for.

Health and Safety documentation will carry some weight if they are generic documents, but if the adjuster can establish that they are bespoke or customised to the company's operations and show evidence that they are accepted and observed by employees/crew, it will go a long way to defending the firework operator or reducing the amount of damages awarded.

Settlement/repudiation

The time a claim can take to settle varies hugely. It would be true to say that the shortest time span for claims are where the loss adjuster repudiates the claim, denying liability, and this is accepted by the third party or their legal representatives. Claims involving damage to property can settle relatively quickly, as the basis of settlement is cost of repair or replacement. Once the question of liability is answered, settlement can be achieved within a few months as long there are no complications or other factors to consider.

However, injury claims other than very minor injuries usually take at least 12 months to settle and can take as much as three or four years. There are both cynical and practical reasons why claims take such long periods to settle. The cynical reason is that claimants and those representing them might hope that recollection of circumstances and physical evidence might be lost or disposed of over a few years and shift the balance in their favour. The practical reasons are that it might take time to recover from an injury and further time to establish the impact of the injury on a claimant's ability to live their life or the type of job they are able to do. This is particularly true where children are concerned because it can be difficult to determine the impact of an injury on a child until they have finished growing. You should expect liability claims involving physical injury to an adult to take at least 18 months to settle; where the claim involves a child the claim may take several years.

Figure 22.1 – A bewildering array of Maltese multi-break cylinder shells NOT transported according to United Nations recommendations!

Life after a claim

Working in the firework industry involves an element of risk. One can manage and control the risk to a certain degree but no one can control the risk entirely, which means there is always the potential of a claim. It is undeniably disturbing when that claim occurs, but with the correct documented systems of work and insurance in place, the risk of claims will be managed and you will have a process in place to advise you, protect your financial exposure and ensure that you can continue with your business.

Notes

1. Useful information at http://www.sja.org.uk/sja/about-us/latest-news/news-archive/news-stories-from-2008/november/fireworks-first-aid.aspx

Chapter 23 – Conclusions

Tom Smith – Davas Ltd

This book is intended to examine the processes necessary to design and fire a variety of firework displays, from small local events using consumer fireworks to international-scale events that are very complex in both design and practicalities.

It is, of course, incomplete — any text sets the scene at a particular moment in time, and in the pyrotechnic field materials and practice are changing rapidly, if subtly, all the time. It is hoped that future editions will address these omissions; but again, they will be limited to the time of publication. Some attempt has been made to "second guess" some of the future developments.

Handling fireworks is potentially dangerous and injuries or fatalities can be caused if operators don't know what they are doing, or think they know what they are doing "because we've always done it like that" but in reality do not. It is to be hoped that this book will improve the knowledge of all those involved in the firework display industry —to varying degrees, of course.

The publisher and authors would be very pleased to receive comments and suggestions to improve the text in future editions.

Bibliography

The following texts are particularly relevant or interesting to display designers, commissioners and firers. Further details are available from http://www.fd — ee.com

BPA Firers handbook	The manual for professional display firers in the UK. Available from the British Pyrotechnists Association – www.pyro.org.uk
Canadian Firework Manual	See http://www.nrcan.gc.ca/mms – smm/expl – expl/pdf/fireworks_e.pdf
Master of the Ceremonies **Ric Birch**	Pub: Allen & Unwin ISBN: 1 – 74114 – 417 – 5
Fireworks – Principles and Practice – 4th Edition **Ron Lancaster**	Pub: Chemical Publishing ISBN: 978 – 0 – 8206 – 0407 – 7
HSG 123 – "Working Together on Firework Displays" – a handbook for event organisers	Pub: HSE Books ISBN: 978 – 071766 – 196 – 1 Available to download from http://www.hse.gov.uk/pubns/books/hsg123.htm
HSG 124 – Giving Your Own Firework Display	Pub: HSE Books ISBN: 978 – 071766 – 162 – 6 Available to download from http://www.hse.gov.uk/pubns/books/hsg124.htm
Fireworks – The Art, Science & Technique **Takeo Shimizu**	Pub: Pyrotechnica Publications (1981) ISBN: 0 – 929388 – 04 – 6
Hanabi – The Fireworks of Japan	Pub: JICC Inc. (1986) ISBN: 0 – 87040 – 706 – 0
The Pyrotechnist's Treasury **Thomas Kentish**	Pub: Hessner Publishing
A Bibliography of Firework Books **Chris Phillip**	Pub: C Philp (1985) ISBN: 0 – 906795 – 31 – 1
Chemistry of Pyrotechnics 2nd edition **J A Conklin & C J Mocella**	Pub: CRC Press (2010) ISBN: 978 – 1 – 57444 – 740 – 8
Lecture notes for firework displays **Ken & Bonnie Kosanke**	Pub: Journal of Pyrotechnics ISBN: 1 – 889526 – 16 – 9
Illustrated Directory of Pyrotechnics **Ken & Bonnie Kosanke**	Pub: Journal of Pyrotechnics ISBN: 1 – 889526 – 01 – 0
Journal of Pyrotechnics	See http://archives.jpyro.com for electronic access to all past and current papers.

References

Where possible URL links are also shown and are all valid at the date of publication. An updated list of references will be made available at http://www.fd-ee.com.

Index

www.ingramcontent.com/pod-product-compliance
Lightning Source LLC
LaVergne TN
LVHW011528200926
848484LV00001B/191